Proceedings, 2020, Ia ValSe-Food 2019

Proceedings, 2020, 1a ValSe-Food 2019

Editors

Isabel Castanheira
Norma C. Sammán
Loreto Muñoz H.
Claudia Monika Haros

MDPI • Basel • Beijing • Wuhan • Barcelona • Belgrade • Manchester • Tokyo • Cluj • Tianjin

Editors
Isabel Castanheira
National Institute of Health
Doutor Ricardo Jorge
Portugal

Norma C. Sammán
Universidad Nacional de Jujuy
Argentina

Loreto Muñoz H.
Universidad Central de Chile
Chile

Claudia Monika Haros
Instituto de Agroquímica y
Tecnología de Alimentos
(IATA-CSIC)
Spain

Editorial Office
MDPI
St. Alban-Anlage 66
4052 Basel, Switzerland

This is a reprint of articles from the Special Issue published online in the open access journal *Proceedings* (ISSN 2504-3900) (available at: https://www.mdpi.com/2504-3900/53/1).

For citation purposes, cite each article independently as indicated on the article page online and as indicated below:

LastName, A.A.; LastName, B.B.; LastName, C.C. Article Title. *Journal Name* **Year**, *Volume Number*, Page Range.

ISBN 978-3-0365-0368-4 (Hbk)
ISBN 978-3-0365-0369-1 (PDF)

Cover image courtesy of Loreto Muñoz.

Contents

About the Editors ix

Preface to "*Proceedings*, 2020, la ValSe-Food 2019 " xiii

Ritva Repo-Carrasco-Valencia
Nutritional Value and Bioactive Compounds in Andean Ancient Grains
Reprinted from: *Proceedings* **2020**, *53*, 1, doi:10.3390/proceedings2020053001 1

Raquel Selma-Gracia, Claudia Monika Haros and José Moisés Laparra
Potential Beneficial Effects of *Chenopodium quinoa* and *Salvia hispanica* L. in Glucose Homeostasis in Hyperglycemic Mice Model
Reprinted from: *Proceedings* **2020**, *53*, 2, doi:10.3390/proceedings2020053002 7

Raquel Selma-Gracia, José Moisés Laparra Llopis and Claudia Monika Haros
Development of New Starch Formulations for Inclusion in the Dietotherapeutic Treatment of Glycogen Storage Disease
Reprinted from: *Proceedings* **2020**, *53*, 3, doi:10.3390/proceedings2020053003 11

Pablo Mezzatesta, Silvia Farah, Amanda Di Fabio and Raimondo Emilia
Variation of the Nutritional Composition of Quinoa According to the Processing Used
Reprinted from: *Proceedings* **2020**, *53*, 4, doi:10.3390/proceedings2020053004 15

Patricia Miranda Villa, Natalia Cervilla, Romina Mufari, Antonella Bergesse and Edgardo Calandri
Making Nutritious Gluten-Free Foods from Quinoa Seeds and Its Flours
Reprinted from: *Proceedings* **2020**, *53*, 5, doi:10.3390/proceedings2020053005 21

Jehannara Calle, Yaiza Benavent-Gil and Cristina M. Rosell
Influence of the Use of Hydrocolloids in the Development of Gluten-Free Breads from *Colocasia esculenta* Flour
Reprinted from: *Proceedings* **2020**, *53*, 6, doi:10.3390/proceedings2020053006 31

María Alejandra Giménez, Cristina Noemí Segundo, Manuel Oscar Lobo and Norma Cristina Sammán
Physicochemical and Techno-Functional Characterization of Native Corn Reintroduced in the Andean Zone of Jujuy, Argentina
Reprinted from: *Proceedings* **2020**, *53*, 7, doi:10.3390/proceedings2020053007 35

Alejandra Chinchilla, Susana Rubio-Arraez, Marisa L. Castelló and María Dolores Ortolá
Enrichment of Protein and Antioxidants of Cupcake with Moringa (*Moringa oleifera*) Leaf Powder and Sensorial Acceptability
Reprinted from: *Proceedings* **2020**, *53*, 8, doi:10.3390/proceedings2020053008 43

Lourdes Wiszovaty, Silvia Caballero, Cristian Oviedo, Fernanda Ozuna and Laura Mereles
Plinia peruviana "Yvapurũ" Fruits and Marmalade from Paraguay: Autochthon Products with Antioxidant Potential
Reprinted from: *Proceedings* **2020**, *53*, 9, doi:10.3390/proceedings2020053009 49

Coronel Eva, Caballero Silvia, Baez Rocio, Villalba Rocio and Mereles Laura
Sicana odorifera "Kurugua" from Paraguay, Composition and Antioxidant Potential of Interest for the Food Industry
Reprinted from: *Proceedings* **2020**, *53*, 10, doi:10.3390/proceedings2020053010 55

Rafael Alarcón, Billy Gonzales, Axel Sotelo, Gabriela Gallardo, María del Carmen Pérez-Camino and Nancy Chasquibol
Microencapsulation of Sacha Inchi (*Plukenetia huayllabambana*) Oil by Spray Drying with Camu Camu (*Myrciaria dubia* (H.B.K.) Mc Vaugh) and Mango (*Mangifera indica*) Skins
Reprinted from: *Proceedings* **2020**, *53*, 11, doi:10.3390/proceedings2020053011 . . . . **61**

Julio Rueda, Manuel Oscar Lobo and Nornma Cristina Sammán
Changes in the Antioxidant Activity of Peptides Released during the Hydrolysis of Quinoa (*Chenopodium quinoa* willd) Protein Concentrate
Reprinted from: *Proceedings* **2020**, *53*, 12, doi:10.3390/proceedings2020053012 . . . . **67**

María Dolores Jiménez, Manuel Oscar Lobo and Norma Cristina Sammán
Technological and Sensory Properties of Baby Purees Formulated with Andean Grains and Dried with Different Methods
Reprinted from: *Proceedings* **2020**, *53*, 13, doi:10.3390/proceedings2020053013 . . . . **75**

Francisco Teodoro Rios, Argentina Angelica Amaya, Manuel Oscar Lobo and Norma Cristina Samman
Design and Acceptability of a Multi-Ingredients Snack Bar Employing Regional PRODUCTS with High Nutritional Value
Reprinted from: *Proceedings* **2020**, *53*, 14, doi:10.3390/proceedings2020053014 . . . . **83**

Ritva Repo-Carrasco-Valencia, Julio Vidaurre-Ruiz and Genny Isabel Luna-Mercado
Development of Gluten-Free Breads Using Andean Native Grains Quinoa, Kañiwa, Kiwicha and Tarwi
Reprinted from: *Proceedings* **2020**, *53*, 15, doi:10.3390/proceedings2020053015 . . . . **91**

Elsa Julieta Salazar de Ariza, Ana Ruth Belloso Archila, Ingrid Odete Sanabria Solchaga and Sandra Beatriz Morales Pérez
Nutritional Composition and Uses of Chia (*Salvia hispanica*) in Guatemala
Reprinted from: *Proceedings* **2020**, *53*, 16, doi:10.3390/proceedings2020053016 . . . . **97**

Talía Hernández-Pérez, María Elena Valverde, Domancar Orona-Tamayo and Octavio Paredes-Lopez
Chia (*Salvia hispanica*): Nutraceutical Properties and Therapeutic Applications
Reprinted from: *Proceedings* **2020**, *53*, 17, doi:10.3390/proceedings2020053017 . . . . **103**

Armando M. Martín Ortega and Maira Rubí Segura Campos
Effect of Chia Seed Oil (*Salvia hispanica* L.) on Cell Viability in Breast Cancer Cell MCF-7
Reprinted from: *Proceedings* **2020**, *53*, 18, doi:10.3390/proceedings2020053018 . . . . **109**

María Gabriela Bordón, Noelia P. X. Alasino, Maria Victoria Defaín Tesoriero, Nahuel Camacho, Maria C. Penci and Marcela L. Martínez and Pablo D. Ribotta
Spray-Air Contact in Tall and Short-Type Spray Dryers Affects Important Physicochemical Properties of Microencapsulated Chia Seed Oil
Reprinted from: *Proceedings* **2020**, *53*, 19, doi:10.3390/proceedings2020053019 . . . . **115**

Luciana M. Julio, Vanesa Y. Ixtaina and Mabel C. Tomás
Development and Characterization of Functional O/W Emulsions with Chia Seed (*Salvia hispanica* L.) by-Products
Reprinted from: *Proceedings* **2020**, *53*, 20, doi:10.3390/proceedings2020053020 . . . . **123**

Natalia Vera, Laura Laguna, Liliana Zura and Loreto A. Muñoz
A Comparative Study of the Physical Changes of Two Soluble Fibers during In Vitro Digestion
Reprinted from: *Proceedings* **2020**, *53*, 21, doi:10.3390/proceedings2020053021 . . . . **129**

About the Editors

Isabel Castanheira, PhD (ORCID: 0000-0001-7273-0676), Principal Researcher, Head of the Food and Nutrition Department, National Institute of Health Doutor Ricardo Jorge.

As Head of Department, she is responsible for consolidating INSA's national and international reputation for developing and applying analytical methods for assessing food quality and food safety. She coordinates activities on the content of nutrients and contaminants of emerging interest in food products in terms of the comparability and reliability of the measurement values. She coordinates the management of activities related to the food–disease relationship. As a member of the WHO Collaborating Centre for Nutrition and Childhood Obesity, she works under WHO recommendations to characterize infant foods and other commodities with relevance to nutritional epidemiology. As a chairperson of the IMEKO Technical Committee for Metrology in Food and Nutrition (2009–2020), she worked on aspects of the traceability routes to SI units involving intermediate reference points which are identified and developed through state-of-the-art analytical methods. She has been a member of international collaborative studies focusing on the certification campaign of values assigned to certified reference materials. She permanently supervises Master's and PhD theses involving: 1) analytical methods; 2) nutrient profile; 3) quality tools applied to food science and technology; 4) inorganic contaminants; 5) speciation analysis; 6) customer perception of label claims; and 7) nutrient losses and gains and yield factors. Research member of several European research projects. She dedicates part of her research activities to the nutritional characterization of traditional foods from Latin America with an impact on health. A reviewer of scientific articles in indexed journals, mainly in the first quartile (SCImago Journal Rank). Associated Editor of Food Chemistry (Elsevier). Member of the editorial boards of the Journal of Food Composition and Analysis (Elsevier), Measurement (Elsevier) and Foods (MPDI). Author of more than 100 scientific publications in various areas, from metrology to food toxicology, with a high share of publications in journals of the first quartile of importance (SCImago Journal Rank).

She has been invited to give lectures in several international food graduate courses under FAO and WHO patronage and at universities around the world. She was chair of 13th International Food Data Conference organized by INFOODS (FAO) held in Lisbon.

Norma C. Sammán, Chemical Engineer, Doctor of Food Technology, Full Professor at the National Universities of Jujuy and Tucumán. She has extensive experience in the chemical, nutritional and functional characterization of foods and the development of new food products. She has made important achievements in the academic, research and transfer fields.

Since 1999, she has been a member of the Regional Academic Commission of the Doctorate in Food Science and Technology of the Northwest of Argentina. She has contributed to the training of human resources through the development of national and international courses and in the direction of more than 20 doctoral theses. She was Director of the Jujuy Research and Transfer Center (CIT-Jujuy) Universidad Nacional de Jujuy (UNJu)-Consejo Nacional de Investigaciones Científicas y Técnicas (CONICET) (2012–2018).

She was the President (2006–2009; 2015–2018) and is now the current Geographic Representative (2018–2021) for South America of LATINFOODS (Latin American Food Composition Network). She has contributed to the development of capacities and standards in food composition. She is a member of the INFOODS Task Force.

Currently, her research activities are oriented to the study of Andean crops in the Argentine Northwest with the main objective of promoting sustainable productive activities. The research group that she directs carries out activities related to the following: i) Survey of the productive and economic-social characteristics of agricultural producers to detect potentialities, making organizational proposals, many of them accepted by the communities involved and compatible with their worldview, values, culture and history. ii) Preservation of genetic material and characterization of the different varieties of Andean potatoes and corn. This material was distributed to agricultural cooperatives to encourage its cultivation and conservation. iii) Development of low-cost and environmentally compatible technologies for the elaboration of new food products using native raw materials that allow high retention of nutrients, biofunctional compounds and their sensory characteristics.

She has published numerous scientific articles, book chapters and technical or policy guidance documents and has received awards for her work in the field of food and nutrition.

Loreto A. Munoz, Food Process Engineer from the Universidad de Santiago de Chile (Chile), Master's in Food Science from Universidad de Chile (Chile), PhD in Science and Food Engineering from Universidad de Santiago de Compostela (Spain) and PhD in Engineering Science from Pontificia Universidad Catolica de Chile (Chile). At present, she is full professor at the Universidad Central de Chile and Head of the Food Science Lab (LabCial).

She has contributed to the training of advanced human capital through the direction of undergraduate and graduate theses.

She has participated and directed numerous projects of national and international scientific and academic significance, as well as significant impact, through ISI scientific publications. In 2013, she received the BIMBO Panamericano Award for her research on the chia seed.

Currently, her research has been focused on the extraction and characterization of food materials of plant origin of interest for human consumption such as seeds, grains and legumes in terms of their microstructural and morphological characterization and nutritional, physical (optical, rheological, textural and thermal properties), chemical and functional properties, as well as studying the digestibility, bioaccessibility and bioavailability of nutrients. In addition, she has experience in the extraction, separation and concentration of their components, as well as their application in food matrices with functional properties.

Finally, her current projects are related to the evaluation of new sources of dietary fiber in terms of the potential contribution on the reduction in risks associated with metabolic syndrome.

Claudia Monika Haros, Chemist, graduated from the School of Exact and Natural Sciences, University of Buenos Aires (UBA), Argentina, in 1990. She also gained her MSc in Bromatology and Food Technology (1992) and MSc in Biology Analysis (1997) from the UBA. She has a PhD in Chemistry (UBA—1999), officially approved by the Ministry of Education, Culture and Sport of Spain. From 1991 to 2003, she worked as university professor in the Organic Chemistry Department, Food Science and Technology Area, UBA. From 1991 to 1999, she was a Research Assistant in the Cereals and Oilseeds Group, Department of Industrial Chemistry, UBA. Later, from 2000 to 2002, she worked in Spain as a visiting professor in the Cereal Group of the Institute of Agrochemistry and Food Technology (IATA) in Valencia. During 2003, she was a postdoc fellow at the Department of Food Microbiology, Institute of Animal Reproduction and Food Research (CENEXFOOD-EU), Polish

Academy of Science, Olsztyn, Poland. From 2003 to 2004, she received an award for working with Prof. Sandberg of the Department of Chemical and Biological Engineering, Life Science Division, University of Chalmers, Gothenburg, Sweden. In 2005, she became a Research Associate (Ramon y Cajal Contract) of the Spanish Council for Scientific Research of the Ministry of Economy and Competitiveness (CSIC-MINECO). Since 2008, she has been a permanent staff member of CSIC and continues her investigation in the Cereal Group, Department of Food Science of IATA. Since 2015, she has been a coordinator of the International Chia-Link Network, and since 2018, she has been the coordinator of la ValSe-Food Group-CYTED.

Since the early stages of her career, she has mainly been engaged in research in respect to the cereal science and technology field. The major theme in Dr. Haros's research is the utilization of different strategies to improve the nutritional and/or functional value of cereal by-products or cereal ingredients. These strategies include the use of different physical, biochemical or biological treatments during the cereal milling process; the development of new cereal by-products by including novel ingredients; and the use of new starter phytase producers for regulating the content and composition of lower myo-inositol phosphates in cereal by-products with clear nutritional and health benefits. In recent years, her research has focused on nutritional studies of raw vegetable materials and/or their by-products to determine their biological activity for their subsequent integration into new food matrices. For this purpose, different in vitro and in vivo strategies are utilized for determining nutritional and/or biological activities. These assays include the determination of the bioaccessibility/bioavailability of minerals, glycaemic index and nutrient inputs according to Dietary Reference Intakes/Adequate Intakes (DRIs/AIs). The ultimate objective is to identify dietetic solutions and innovations to prevent diseases and to improve consumers' well-being/health.

https://www.iata.csic.es/en/staff/claudia-monika-haros
https://publons.com/researcher/1751589/monika-haros/
ID: H-6839-2012
http://orcid.org/0000-0001-7904-0109

The editors belong to the Chia-Link Network and la ValSe-Food CYTED Network
http://www.chialink.es/
http://www.cyted.org/es/valse_food

Preface to "*Proceedings*, 2020, la ValSe-Food 2019 "

Iberoamerican crops are underutilized and cultivation levels are low, but, recently, their worldwide demand has significantly increased, resulting in a production increase as well as a price increase. For many years, these valuable seeds have been widely recognized by food scientists and food producers because of their nutritional value. They contain high-quality proteins, and some contain abundant amounts of starch and/or fiber (with unique characteristics) and large quantities of micronutrients such as minerals, vitamins and bioactive compounds; moreover, they are gluten-free, which makes them suitable for people suffering from gluten intolerances/allergies. For these reasons, the Iberoamerican valuable seeds interest has immensely increased, since in recent decades, the research efforts have been intensified.

This book summarizes the Proceedings of the II International la ValSe-Food, Development of Food Ingredients from Iberoamerican Ancestral Crops and V Symposium of Chia-Kink Network held at the National Institute of Health Doutor Ricardo Jorge (INSA), 20–21 October, Lisbon, Portugal. This book gathers the recent investigations of la ValSe-Food Group on these valuable seeds and other crops and provides comprehensive and up-to-date knowledge within all the relevant fields of food science. It provides information on production and utilization, structure and chemical composition, paying special attention to carbohydrates, fibers, bioactive compounds, proteins and lipids of kernels and other parts of the plants. It includes their processing, various food products and applications and the nutritional and health implications.

We hope that this book will contribute to the increased utilization of Iberoamerican valuable seeds in human nutrition.

This publication was financially supported by Ia ValSe-Food-CYTED (119RT0567).

Isabel Castanheira, Norma C. Sammán, Loreto Muñoz H., Claudia Monika Haros
Editors

proceedings

Proceedings

Nutritional Value and Bioactive Compounds in Andean Ancient Grains †

Ritva Repo-Carrasco-Valencia

Centro de Investigación e Innovación en Productos Derivados de Cultivos Andinos CIINCA, Universidad Nacional Agraria La Molina, Avenida La Molina s/n Lima 18, Lima 15024, Peru; ritva@lamolina.edu.pe

† Presented at the 2nd International Conference of Ia ValSe-Food Network, Lisbon, Portugal, 21–22 October 2019.

Published: 3 August 2020

Abstract: Quinoa (*Cheopodium quinoa*), kañiwa (*Cheopodium pallidicaule*), kiwicha (*Amaranthus caudatus*) and tarwi (*Lupinus mutabilis*) are ancient crops from the Andean region of South America. Recently, interest in these crops has grown, and worldwide demand for them has increased considerably. The aim of this study was to study the bioactive compounds and nutritional compositions of different varieties/ecotypes of quinoa, kañiwa, kiwicha and tarwi. Proximate, mineral, dietary fibre, fatty acid and amino acid compositions were evaluated. The content of phenolic compounds, tocopherols and phytosterols, and the folic acid and antioxidant capacity, were determined as well. The protein content of the grains was between 13.00% and 20.00%. More important than protein quantity is protein quality, which is demonstrated by the composition of the amino acids. All analysed grains, and especially the kañiwa, had very high lysine content. This amino acid is especially important in vegetarian diets because it is the limiting amino acid in cereal protein. The content of the total phenolic compounds in the studied grains was 27–58 mg gallic acid/100 g of sample. In quinoa, the principal flavonoids found were quercetin and kaempferol, in kañiwa quercetin and isorhamnetin. In kiwicha, no detectable amounts of flavonoids were found. Plant sterols (phytosterols) were another group of biologically active compounds detected. Andean lupin, tarwi, is very rich in oil, which has a beneficial nutritional composition. In conclusion, all studied grains have a very high nutritional value, are interesting sources of bioactive compounds and could be used as ingredients in health-promoting functional foods.

Keywords: quinoa; kañiwa; kiwicha; tarwi

---

1. Introduction

The Andean region of South America is an important center of the domestication of food crops. This area has a diversity of landscapes and agroecological zones, due to several climates and altitude differences (1500–4200 m). Compared with other regions in the world where crops have been domesticated, the Andean region has its own characteristics. There are no vast, unending plains of uniformly fertile, well-watered land, as in Asia, Europe and the Middle East. Instead, there is an almost total lack of flat, fertile, well-watered soil. The Andean people have always cultivated their crops on tiny plots, one above another up mountain sides, rising thousands of meters [1].

At the time of the European conquest, the Incas cultivated almost as many species of plants as the farmers of all Asia or Europe. It has been estimated that Andean native people domesticated as many as 70 separate crop species [1]. On mountain sides up to four km high along the whole continent, and in climates varying from tropical to polar, they grew roots, grains, legumes, vegetables, fruits and nuts.

Andean indigenous grains, such as quinoa (*Chenopodium quinoa*), kañiwa (*Chenopodium pallidicaule),* kiwicha (*Amaranthus caudatus*) and tarwi (*Lupinus mutabilis),* are good sources of high-quality proteins. They contain also dietary fibre and oil with polyunsaturated fatty acids. Dietary fibre is especially important in diets designated for disease risk reduction and the prevention of diabetes and heart disease. Quinoa, kañiwa and kiwicha are sometimes called pseudocereals because of their similarity in chemical composition with common cereals, such as wheat and rice. Tarwi is a leguminous seed grown mainly in Andean highlands.

The Andean indigenous food crops have enormous potential to be used as functional foods in the prevention of chronic diseases, such as cardiovascular diseases, cancer and diabetes. High variability, not only in colours and shapes, but also in primary nutrient constituents and bioactive compounds, has recently been reported. The health-related properties of Andean crops claimed by local people could be partially attributed to the presence of these bioactive compounds.

The objective of this research was to study the chemical composition and some bioactive compounds of different varieties/ecotypes of quinoa, kañiwa, kiwicha and tarwi.

2. Materials and Methods

Two varieties of quinoa and kiwicha and one variety of tarwi and kañiwa were acquired from the southern Andes of Peru.

Water content, proteins (N × 6.25), fat, crude fibre and ashes were determined according to American Association of Cereal Chemists, AACC (2005) [2].

The total dietary fibre was analysed by an enzymatic-gravimetric method according to the Approved Method of AACC (2005) [2] using the TDF-100 kit from Sigma Chemical Company (St. Louis, MO, USA)

Radical scavenging activity was determined according to the method of Brand-Williams et al. [3] (217) based on the decrease of absorbance at 515 nm produced by reduction of DPPH (2, 2Diphenyl-1-picrylhydrazyl) by an antioxidant. Trolox was used as the reference compound.

The content of total phenolics was analysed according to the method of Swain and Hillis (220).

Tocopherols and tocotrienols were determined based on the direct hydrolysis method reported by Fratianni et al. [4].

The fatty acid composition of the fat fraction was determined after methylation using a modification of the procedure described by Slover and Lanza [5]. Profiling analysis of fatty acid methyl esters was conducted on a 6890N GC-FID gas chromatograph (Agilent) equipped with an Omegawax 250 fused silica capillary column (30 m × 0.25 mm × 0.25 μm, Supelco Inc., Bellefonte, PA, USA). The initial oven temperature was held at 160 °C for 1 min, raised to 240 °C at a rate of 4 °C/min, and kept there for 5 min. The injector and detector temperatures were 240 °C and 260 °C, respectively. Helium was used as the carrier gas at a flow rate of 1.1 mL/min. Identification of fatty acid methyl esters was carried out by comparing their retention times with those of standards (Sigma, USA). Results were expressed as percentage of total fatty acid methyl esters analysed.

Mineral: Ca, Mg, P, Fe. Samples were digested in concentrated nitric acid in a Tecator block digestor. ICP-OES = Inductively Coupled Plasma-Optical Emission Spectrometry was used for the determination of mineral and trace elements [6].

3. Results and Discussions

The proximate composition of the grains is presented in Table 1.

Table 1. Chemical composition of Andean grains (g/100 g).

Samples	Moisture	Protein	Fat	Ash	Dietary Fibre
Quinoa 1	10.9 ± 0.08	17.9 ± 0.11	7.0 ± 0.03	3.1 ± 0.02	12.4 ± 0.57
Quinoa 2	10.2 ± 0.09	15.3 ± 0.09	7.4 ± 0.04	3.3 ± 0.03	18.1 ± 0.81
Kañiwa	11.2 ± 0.08	17.0 ± 0.11	8.9 ± 0.02	3.8 ± 0.08	20.9 ± 0.46
Kiwicha 1	9.8 ± 0.04	14.8 ± 0.11	8.3 ± 0.03	2.2 ± 0.02	9.0 ± 0.30
Kiwicha 2	9.2 ± 0.07	15.0 ± 0.00	7.6 ± 0.05	3.0 ± 0.05	16.4 ± 0.55
Tarwi	8.2 ± 0.06	36.9 ± 0.15	18.4 ± 0.06	3.6 ± 0.05	37.0 ± 0.85

Analysis was made as triplicates.

The protein content of the three Andean grains was between 15.0% and 17.9%. The tarwi had a very high protein content (36.9%). The values of the proteins detected in this study are similar to the values found by Repo-Carrasco-Valencia [7]. The differences between the samples in protein content seem to be more related to the place of cultivation than to variety. The plants use the nitrogen from the soil to produce the proteins. If the soil is rich in nitrogen, the plant produces more proteins than a plant which has been cultivated in nitrogen-poor soil.

In the comparison of the proximate compositions of the Andean grains with the proximate compositions of common cereals, we can find some similarities and differences. The content of total carbohydrates in cereals and Andean grains is similar, about 60–75%. The main carbohydrate is starch in all grains. Thus, they could be used as materials for starch industries. The fat content in Andean grains is considerably higher than the fat content in common cereals (6–7% vs. 2–4%) [8]. The oil is of good nutritional quality, containing the essential fatty acids in proper proportions (see Table 2). Quinoa, kañiwa, kiwicha and tarwi could serve as raw materials to produce edible oils.

Table 2. Fatty acid composition of Andean grains (% of total fatty acids).

Sample	Myristic C14:0	Palmitic C16:0	Stearic C18:0	Oleic C18:1	Linoleic C18:2w6	Linolenic C18:3w3
Quinua	0.14	10.08	0.66	26.15	46.83	8.21
Quinua	0.10	9.1	0.79	27.73	50.38	4.37
Kañiwa	0.12	12.89	1.31	24.86	48.37	5.25
Kiwicha	0.17	17.01	3.01	29.04	42.55	1.84
Kiwicha	0.15	16.64	3.84	30.27	42.17	0.83
Tarwi	n.d.	9.60	9.06	51.21	24.23	2.76

In Table 3, the contents of phenolic compounds and the antioxidant capacities of Andean grains can be observed. The kañiwa had the highest value compared with the other grains, followed by tarwi, and the kiwicha samples had the lowest values. This sample tendency can be observed in the values of antioxidant capacity. This demonstrates that the phenolic compounds are the main compounds responsible of the antioxidant activity of these Andean grains.

Tocopherols are compounds with a high antioxidant capacity and other important physiological functions. Some of them have the function of vitamin E. The contents of different tocopherols in the Andean grains are presented in Table 4. The highest content of tocopherols was found in black quinoa from Cusco. Kiwicha samples were low in α-tocopherol, but interestingly, high in γ-tocopherol. This tocopherol was found in high amounts in kañiwa as well. Wheat has about 1.3 mg/100 g of total α-tocopherol, and it does not contain γ-tocopherol [9]. Barley, oat, rye, rice and corn contain the following amounts of total tocopherols: 0.75–0.9, 0.6–1.3, 1.8, 0.2–0.6 and 4.4–5.1 mg/100 g, respectively [8]. Thus, in comparison with common cereals, the Andean grains could be considered good sources of these tocopherols.

Table 3. Phenolic compounds and antioxidant capacity in Andean grains.

Sample	Total Phenolics (mg Gallic Acid/100g of Sample)	Antioxidant Capacity with DPPH (µg of Trolox/g of Sample)
Quinoa 1	34.0	388.2
Quinoa 2	40.6	654.0
Kañiwa	57.9	1566.8
Kiwicha 1	17.4	172.7
Kiwicha 2	15.5	202.0
Tarwi	45.0	620.0

Table 4. Content of tocopherols in Andean grains.

Sample	α-Tocopherol mg/100 g dw	γ-Tocopherol mg/100 g dw	δ-Tocopherol mg/100 g dw
Quinoa 1	4.1	3.7	<0.2
Quinoa 2	2.3	5.8	0.6
Kañiwa	1.6	6.0	<0.2
Kiwicha 1	0.5	4.7	2.7
Kiwicha 2	0.6	3.9	2.0

The contents of some important minerals in Andean grains can be observed in Table 5. In comparison with quinoa and kañiwa samples, the kiwicha samples had the highest calcium content. The black kiwicha was exceptionally rich in this element. This sample had the highest iron and magnesium content, as well. The content of calcium and magnesium in Andean grains is higher than the content of these minerals in wheat, barley, oats and rice, whereas the content of iron is similar [9].

Table 5. Content of minerals in Andean grains.

Sample	Calcium (g/kg dw)	Magnesium (g/kg dw)	Iron (mg/kg dw)
Quinua 1	0.54	2.13	55
Quinua 2	0.49	2.29	50
Kañiwa	1.21	2.95	57
Kiwicha 1	1.37	2.11	54
Kiwicha 2	3.06	3.38	73

4. Conclusions

All Andean grains had a relatively high protein and fat content. The pink kiwicha had the highest oil content. This is interesting because kiwicha oil is a very good source of phytosterols, essential fatty acids and, according to the literature, of squalene. This compound is found in olive oil, as well. Squalene is a terpenoid compound and it has some health-promoting properties; in particular it lowers the cholesterol level in blood by inhibiting its synthesis in the liver. In addition, it is hypothesized that the decreased risk of various cancers associated with high olive oil consumption could be due to the presences of squalene. This variety could be an interesting material for producing edible high-quality oil, with bioactive compounds, such as squalene, essential fatty acids and phytosterols. Kañiwa's oil is very interesting, as well. It has a very high content of tocopherols and unsaturated fatty acids.

The best source of antioxidants (phenolic compounds) out of the studied grains was kañiwa. Andean grains contain flavonoids, a type of phenolic compound with important antioxidant activity. Berries, such as blueberries, are generally considered to be excellent sources of flavonoids. Regarding the content of tocopherols, in comparison with common cereals, the Andean grains could be considered good sources of these compounds. In general, these Andean native grains are very rich in health-promoting compounds, and should be explored in future studies on the bioavailability of these compounds.

Acknowledgments: This work was supported by grant IaValSe-Food-CYTED (Ref. 119RT0567), International Trade Centre (ITC), Geneva, Switzerland and PROTEIN2FOOD Project (European Union's Horizon 2020, No. 635727).

References

1. National Research Council. *The Crops of the Incas: Little-Known Plants of the Andes with Promise for Worldwide Cultivation*; National Academy Press: Washington, DC, USA, 1989.
2. AACC Approved Methods. *American Association of Cereal Chemists*, 10th ed.; Cereals and Grains Association: St. Paul, MN, USA, 2005.
3. Brand-Willians, W.; Cuvelier, M.; Berset, C. Use of a free radical method to evaluate antioxidant activity. *LWT Food Sci. Technol.* **1995**, *28*, 25–30, doi:10.1016/S0023-6438(95)80008-5.
4. Fratianni, A.; Caboni, M.F.; Irano, M.; Panfili, G. A critical comparison between traditional methods and supercritical carbon dioxide extraction for the determination of tocochromanols in cereals. *Eur. Food Res. Tech.* **2002**, *215*, 353–358, doi:10.1007/s00217-002-0566-2.
5. Slover, H.; Lanza, E. Quantitative analysis of food fatty acids by capillary gas chromatography. *J. Am. Oil Chem. Soc.* **1979**, *56*, 933–943, doi:10.1007/BF02678823.
6. Mattila, P.; Mäkinen, S.; Eurola, M.; Jalava, T.; Pihlava, J.M.; Hellström, J.; Pihlanto, A. Nutritional value of commercial protein-rich plant products. *Plant Foods Hum. Nutr.* **2018**, *73*, 108–115, doi:2010.1007/s11130-018-0660-7.
7. Repo-Carrasco-Valencia, R. Andean Indigenous Crops: Nutritional Value and Bioactive Compounds. Ph.D. Thesis, Department of Biochemistry and Food Chemistry, University of Turku, Turku, Finland, 2011.
8. Kent, N. *Technology of Cereals*; Pergamon Press: Oxford, UK, 1983.
9. Bock, M.A. Minor constituents of cereals. In *Handbook of Cereal Science and Technology*, 2nd ed.; Kulp, K., Ponte, J.G., Eds.; Marcel Dekker: New York, NY, USA; Basel, Switzerland, 2000; p. 790.

proceedings

MDPI

Proceedings

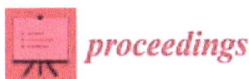

Potential Beneficial Effects of *Chenopodium quinoa* and *Salvia hispanica* L. in Glucose Homeostasis in Hyperglycemic Mice Model [†]

Raquel Selma-Gracia [1,2], Claudia Monika Haros [2] and José Moisés Laparra [1,*]

1 Molecular Immunonutrition Group, Madrid Institute for Advanced Studies in Food (IMDEA-Food), Ctra. de Canto Blanco n° 8, 28049 Madrid, Spain; raquelselgra@iata.csic.es
2 Instituto de Agroquímica y Tecnología de Alimentos (IATA), Consejo Superior de Investigaciones Científicas (CSIC), Av. Agustín Escardino 7, Parque Científico, 46980 Paterna, Valencia, Spain; cmharos@iata.csic.es
* Correspondence: moises.laparra@imdea.org
† Presented at the 2nd International Conference of Ia ValSe-Food Network, Lisbon, Portugal, 21–22 October 2019.

Published: 3 August 2020

Abstract: Impaired glucose homeostasis is associated with an increased risk of developing metabolic alterations. In this study, a model in which mice treated with streptozotocin were fed a high-fat diet was used to mimic early stages of the onset of metabolic disorders, and different bread formulations were administrated to evaluate the effect of replacing wheat flour with *Chenopodium quinoa* (Q) (25%) and *Salvia hispanica* L. (Sh) (20%). Plasmatic glucose and insulin concentrations were quantified and the homeostasis model assessment (HOMAir) was calculated. Q and Sh showed a lower tendency to hyperglycemia compared to wheat bread (WB). Besides, these low glucose levels were accompanied by three-fold lower values of HOMAir respect to WB, suggesting an improved insulin sensitivity. Thus, inclusion of *C. quinoa* and *S. hispanica* into bread formulations could improve the control of glucose homeostasis, which could help to prevent/ameliorate metabolic glucose alterations.

Keywords: *Chenopodium quinoa*; *Salvia hispanica* L.; hyperglycemia; HOMAir

1. Introduction

A good control in glucose homeostasis has shown to be determinant in the prevention and improvement of metabolic disorders, such as insulin resistance, obesity, or type 2 diabetes (T2D) [1]. In recent years, Latin-American crops have become a focus of study because of their beneficial nutritional properties [2]. *Chenopodium quinoa* and *Salvia hispanica* L. in bread formulations [3] have shown an enhanced insulin signaling modulating peroxisome proliferator-activated receptor (PPAR)-γ activation, which has been related to a marked improvement of whole-body insulin sensitivity in T2D patients [2,4].

Thus, the objective of the present investigation was to evaluate the inclusion of flour from *C. quinoa* and *S. hispanica* as a substitute to wheat flour in bread formulations in the control of glucose homeostasis in a hyperglycemic-mouse model.

2. Materials and Methods

2.1. Sample Breads

White quinoa seeds (Organic quinoa Real©) from ANAPQUI, La Paz, Bolivia, were purchased from Ekologikoak (Bizkaia, Spain). Chia flour was purchased from the Primaria Premium Raw Materials Company (Valencia, Spain). Wheat flour was provided from a local market. Three bread formulations were prepared: white quinoa (Q) at 25% and chia (Sh) at 20% and were compared to wheat bread (WB), as a control [3].

2.2. Experiment Design

Female C57BL/6 mice were obtained from Centro de Investigaciones Biológicas (CIB-CSIC) in Madrid, Spain. All animals were injected intraperitoneally with 2 low doses of streptozotocin (25 mg/kg) and they received a high-fat diet for 3 weeks. Animals were distributed into groups depending on the administered bread, and they received 3 doses (14 mg/day/animal) per week during all the treatment.

2.3. Glucose and Insulin Concentration

Insulin and glucose concentrations were determined in plasma samples. Insulin was measured by ELISA kit (RAB0817-1KT, Sigma-Aldrich, Darmstadt, Germany) and glucose (MAK263-1KT, Sigma-Aldrich, Darmstadt, Germany).

2.4. Homeostatic Model Assessment of Insulin Resistance (HOMAir)

Insulin resistance (IR) was defined as the HOMAir value and was calculated according to the following formula: [insulin (μU/mL) × glycaemia (mg/dL)]/405 [5].

2.5. Statistical Analysis

Multiple ANOVA and Fisher's least significant differences (LSD) were applied to establish statistically significant differences. Statistical analyses were performed with the software Statgraphics Centurion XVI and the significance level was established at $p < 0.05$.

3. Results and Discussion

This study showed that metabolic alterations occurring at early stages, onset, and disease progression derived from impairment of glucose homeostasis, improved with the inclusion of flours obtained from *C. quinoa* and *S. hispanica* into bread formulations. Hyperglycemia is a hallmark of a profoundly altered metabolism, which is associated with stress oxidative and inflammatory response in the liver [1]. In this study, glucose blood concentrations decreased, following this order: WB > Q > Sh (Figure 1), displaying a positive effect in Q and Sh and suggesting a more controlled glucose homeostasis, which could have important consequences limiting the hepatic endogenous glucose output and liver metabolic stress [6]. The glycemic index is used to determinate the total rise in blood glucose level; however, it could not reflect the real absorption and modulation in glucose homeostasis. The latter is supported by the lack of differences in the glycemic index reported between *C. quinoa* and WB and the higher transcript levels of PPAR-γ in *C. quinoa* [2]. These observations are concordant with the low HOMAir value reported in animals fed *C. quinoa* (Figure 1).

HOMAir is considered an index of insulin resistance, which is a risk factor of the progression of type 2 diabetes. Animals administered with WB showed values above 3.8, which is indicative of insulin resistance, meanwhile Q and Sh were below. In line with the glucose levels, Sh obtained 1.5-fold lower HOMAir than Q. The effect of *S. hispanica* seeds in a state of insulin resistance has shown an improved sensitivity associated with increased expression levels of the peroxisome proliferator-activated receptor-γ coactivator-1α (PGC-1α) and promoting the absorption of glucose by muscle in obese rats [7]. In line with these positive effects of flour from S. *hispanica*, it was demonstrated that long-term supplementation with *S. hispanica* attenuates cardiovascular risk factors with the reduction of systolic blood pressure and serum C-reactive protein concentration in diabetic patients [8].

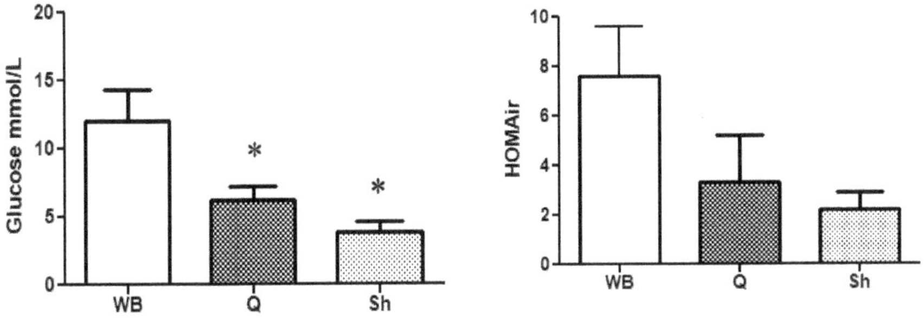

Figure 1. Glucose and HOMAir in mice fed with a high-fat diet and administered with the different bread formulations: WB, white bread; Q, quinoa flour (25%)-containing bread; Sh, chia flour (20%)-containing bread. * Indicates statistically significant ($p < 0.05$) differences in relation to WB.

4. Conclusions

Replacing flour wheat with *S. hispanica* and *C. quinoa* flour could exert beneficial effects, improving insulin resistance and control of glucose homeostasis and, thus, promoting a better metabolic modulation that could prevent the onset or progress of early metabolic alterations.

Funding: This work was financially supported by Ia ValSe-Food-CYTED (119RT0567) and QuiSalhis-Food (AGL2016-75687-C2-1-R) from the Ministry of Economy, Industry and Competitiveness (MEIC).

Acknowledgments: The contract given to R. Selma-Gracia by LINCE (PROMETEO/2017/189) from the Generalitat Valenciana (Spain) is gratefully acknowledged. JML thanks MICINN for his 'Ramon y Cajal' contract (RYC-2015-18083).

References

1. Hamden, K.; Carreau, S.; Boujbiha, M.A.; Lajmi, S.; Aloulou, D.; Kchaou, D.; Elfek, A. Hyperglycaemia, stress oxidant, liver dysfunction and histological changes in diabetic male rat pancreas and liver: Protective effect of 17β-estradiol. *Steroids* **2008**, *73*, 495–501, doi:10.1016/j.steroids.2007.12.026.
2. Laparra, J.M.; Haros, M. Inclusion of whole flour from Latin-american crops into bread formulations as substitute of wheat delays glucose release and uptake. *Plant Foods Hum. Nutr.* **2018**, *73*, 13–17, doi:10.1007/s11130-018-0653-6.
3. Iglesias-Puig, E.; Haros, M. Evaluation of dough and bread performance incorporating chia (*Salvia hispanica* L.). *Eur. Food Res. Technol.* **2013**, *237*, 865–874, doi:10.1007/ s00217-013-2067-x.
4. Kintscher, U.; Law, R.E. PPARγ-mediated insulin sensitization: The importance of fat versus muscle. *Am. J. Physiol. Endocrinol. Metab.* **2005**, *288*, E287–E291, doi:10.1152/ajpendo.00440.2004.
5. Matthews, D.R.; Hosker, J.P.; Rudenski, A.S.; Naylor, B.A.; Treacher, D.F.; Turner, R.C. Homeostasis model assessment: Insulin resistance and *β*-cell function from fasting plasma glucose and insulin concentrations in man. *Diabetologia* **1985**, *28*, 412–419, doi:10.1007/bf00280883.
6. Sharabi, K.; Tavares, C.D.J.; Rines, A.K.; Puigserver, P. Molecular pathophysiology of hepatic glucose production. *Mol. Asp. Med.* **2015**, *46*, 21–33, doi:10.1016/j.mam.2015.09.003.
7. Marineli, R.D.; Moura, C.S.; Moraes, E.A.; Lenquiste, S.A.; Lollo, P.C.B.; Morato, P.N.; Amaya-Farfan, J.; Marostica, M.R. Chia (*Salvia hispanica* L.) enhances HSP, PGC-1α expressions and improves glucose tolerance in diet-induced obese rats. *Nutrition* **2015**, *31*, 740–748, doi:10.1016/j.nut.2014.11.009.

8. Vuksan, V.; Whitham, D.; Sievenpiper, J.L.; Jenkins, A.L.; Rogovik, A.L.; Bazinet, R.P.; Vidgen, E.; Hanna, A. Supplementation of conventional therapy a with the novel grain salba (*Salvia hispanica* L.) improves major and emerging cardiovascular risk factors in type 2 diabetes: Results of a randomized controlled trial. *Diabetes Care* **2007**, *30*, 2804–2810, doi:10.2337/dc07–1144.

proceedings

MDPI

Proceedings

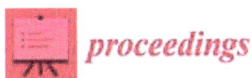

Development of New Starch Formulations for Inclusion in the Dietotherapeutic Treatment of Glycogen Storage Disease †

Raquel Selma-Gracia [1,2], José Moisés Laparra Llopis [2] and Claudia Monika Haros [1,*]

[1] Instituto de Agroquímica y Tecnología de Alimentos (IATA), Consejo Superior de Investigaciones Científicas (CSIC), Av. Agustín Escardino 7, Parque Científico, 46980 Paterna, Valencia, Spain; raquelselgra@iata.csic.es

[2] Molecular Immunonutrition Group, Madrid Institute for Advanced Studies in Food (IMDEA-Food), Ctra. de Canto Blanco n° 8, 28049 Madrid, Spain; moises.laparra@imdea.org

* Correspondence: cmharos@iata.csic.es

† Presented at the 2nd International Conference of Ia ValSe-Food Network, Lisbon, Portugal, 21–22 October 2019.

Published: 4 August 2020

Abstract: In this study, the thermal properties of quinoa and maize starch were evaluated and related to their digestibility. Lower gelatinisation and retrogradation parameters were obtained in quinoa starch, suggesting a better susceptibility to the disruption of the crystalline structure. These results were accompanied with a higher percentage of hydrolysis in raw quinoa, reaching more twofold higher than in raw maize starch. Besides, the slopes calculated by a Lineweaver-Bürke transformation showed similar values in raw quinoa and maize starches. Taken together, these characteristics of quinoa starch could provide more digestible benefits than the current treatment, raw maize starch, in glycogen storage disease patients.

Keywords: maize starch; quinoa starch; thermal properties; starch hydrolysis; glycogen storage disease

---

1. Introduction

Previous research has shown variability in the susceptibility to digestion depending on structural differences in starches from different sources [1]. A crystalline structure is an important factor to take into account in digestibility, and can be modified by a gelatinisation process [2]. These altered structural changes rely on starch type and, as a result, each starch shows a different digestibility [3]. Besides, the extent of digestibility has been known to be related to the degree of polymerisation (DP) of amylopectin, affecting functional properties of the starches, such as thermal parameters, as well as their digestibility [4]. The identification of new sources of starch to obtain an improved digestibility than the current treatment, raw maize starch, could help in minimising digestive and metabolic disturbances for patients with glycogen storage diseases (GSDs).

2. Materials and Methods

2.1. Starches

Commercial maize starch was provided by ACH Food Companies (Argo, IL, USA). Red quinoa starch was obtained in the laboratory by wet milling [5]. Enzymes were purchased from Sigma-Aldrich: α-amylase (EC 3.2.1.1, A3176-1MU, St Louis, MO, USA, 16 U/mg), amyloglucosidase from

Aspergillus niger (EC 3.2.1.3, 10115, Buchs, Switzerland, 60.1 U/mg) and pepsin (EC 3.4.23.1, P7000, Gillingham, UK, 480 U/mg).

2.2. Thermal Properties

Gelatinisation and retrogradation properties were determined using differential scanning calorimetry (DSC) (Perking-Elmer DSC-7, Norwalk, CT, USA). The procedure was done according to the method described by Haros et al. [6] with slight modifications. Water:starch ratio was 3:1 and calorimeter scan conditions were kept at 25 °C for 1 min and then heated from 25 °C to 120 °C at 10 °C/min. To analyse retrograded starch, the samples were stored in a refrigerator for a week and the same process was repeated.

2.3. Preparation of Samples for Digestion

Aliquots (100 mg) of starch samples were weighed and 1 mL of water was added. Raw starches stayed in unheated water and gelatinised starches were kept at 100 °C for 5 min, as a positive control.

2.4. In Vitro Starch Digestion and Glycaemic Index (GI) Estimation

The rate of starch hydrolysis was evaluated according to the method described by Goñi et al. [7] with modifications. Briefly, 10 mL HCl-KCl buffer (pH 1.5) and 400 μL of a solution of pepsin (0.1 g/mL) were added and starches were kept in a shaking water bath at 37 °C for 1 h. Afterwards, 19.6 mL Tris-Maleate buffer (pH 6.9) and 1 mL of a solution containing α-amylase (0.01 g/mL) were added. Aliquots were taken from 0 to 120 min and the enzyme was inactivated. Finally, glucose, area under the curve (AUC), hydrolysis index ((HI) and GI were determined by spectrophotometry according to a commercially available enzymatic kit (D-Glucose Assay Procedure, K-GLUC 07/11, Megazyme) [8]. All the reagents used were analytical grade or better.

2.5. Statistical Analysis

Multiple ANOVA and Fisher's least significant difference (LSD) were carried out for the thermal properties and Tukey's test for the digestion values. The statistical analyses were realised with the software Statgraphics Centurion XVI, and the significance level was established at $p < 0.05$.

3. Results and Discussion

Gelatinisation parameters were determined in maize and quinoa starch, and quinoa displayed lower values than those of maize (Table 1). Thermal parameters have shown a positive correlation with the DP of amylopectin, where maize starch presented a higher DP (12–18) than the quinoa DP (8–10) [4,9]. When retrogradation parameters were measured (4 °C) after 7 days, quinoa starch exhibited the highest resistance to retrogradation, as the short chains of amylopectin in quinoa could have contributed to a lower rearrangement of the starch structure. The proportion of these shorter amylopectin chains in quinoa could have affected the crystalline structure, resulting in a higher susceptibility to enzymatic action [9].

Table 1. Preliminary gelatinisation parameters by differential scanning calorimetry (DSC) *.

Starch	Gelatinisation				Retrogradation			
	To (°C)	Tp (°C)	Tc (°C)	ΔH (J/g)	To (°C)	Tp (°C)	Tc (°C)	ΔH (J/g)
Maize	65 ± 1 [b]	70 ± 1 [b]	76 ± 1 [b]	13 ± 1 [b]	44 ± 1 [b]	54 ± 1 [b]	63 ± 1 [b]	3.5 ± 1 [a]
Quinoa	51 ± 2 [a]	59 ± 1[a]	69 ± 1 [a]	10 ± 1 [a]	37 ± 1[a]	47 ± 1 [a]	56 ± 1 [a]	1.6 ± 1 [a]

[a] Mean ± standard deviation, n = 3. Values in the same column followed by the same letter are not [significantly different (p < 0.05); [b] To: Onset temperature, Tp: Peak temperature, Tc: Conclusion temperature, ΔH: Enthalpy change; Mean ± standard deviation, n = 2. * Selma-Gracia et al. [10].

The digestibility of starch represents an important parameter to consider in the severity and clinical manifestations of GSD. Only 30% of raw maize starch was hydrolysed, unlike raw quinoa, which obtained high proportions of hydrolysis from the beginning, reaching up to around 70% (Figure 1). From the accumulated curves of hydrolysis, the Lineweaver-Bürke transformation [8] was calculated to estimate the kinetics of hydrolysis. For raw starches, close values for the slope of the plotted lines for maize (slope = 2.6 SH/min) and quinoa (slope = 2.2 SH/min) (Table 2) were calculated. Thus, similar slope values, but a higher hydrolysis of raw quinoa, could imply the maintenance of optimal glucose levels for a longer time with quinoa starch compared to maize starch. Besides, raw quinoa starch could help in improving the digestive inconveniences derived from the need to consume high amounts of raw maize starch in patients with GSD. When starches were heat treated, gelatinised samples displayed significant differences between the kinetics of hydrolysis: maize, slope = 11.7 SH/min; quinoa, slope = 3.9 SH/min. These differences were reflected mainly in the higher proportions of hydrolysis calculated for gelatinised maize starch within the first 20 min (Figure 1).

Table 2. Preliminary hydrolysis of starches by α-amylase.

Starch	Treatment	TSH_{120} (%)	AUC	HI (%)	GI	Slope (SH/min)
Maize	Raw	30 ± 3 [a]	2975 ± 323 [a]	39 ± 4 [a]	61 ± 2 [a]	2.6 ± 0.2 [a]
	Gelatinised	73 ± 4 [bc]	7408 ± 572 [c]	97 ± 7 [c]	93 ± 4 [c]	11.7 ± 2.1 [b]
Quinoa	Raw	67 ± 4 [b]	5741 ± 606 [b]	75 ± 8 [b]	81 ± 4 [b]	2.2 ± 0.6 [a]
	Gelatinised	82 ± 2 [c]	7153 ± 167 [c]	94 ± 2 [c]	91 ± 1 [c]	3.9 ± 0.5 [a]

[a] Values in the same column followed by the same letter are not significantly different ($p < 0.05$); [b] TSH_{120}: Total starch hydrolysed at 120 min, AUC: Area under the curve of starch digestion from 0 to 120 min, HI: Hydrolysis index, GI: Glycaemic index; [c] Slope was calculated using the Lineweaver-Bürke transformation.

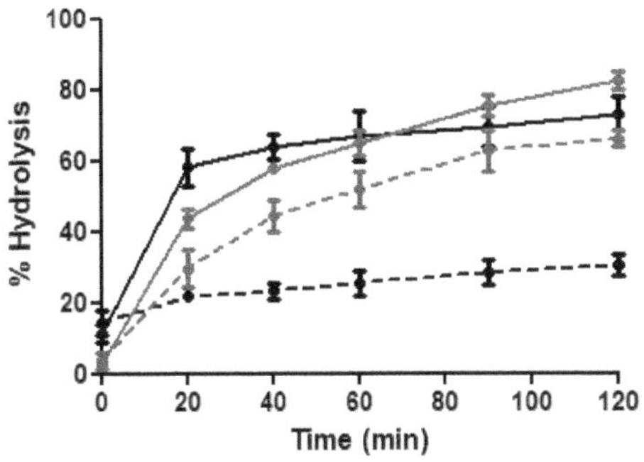

Figure 1. Kinetics of hydrolysis in raw starches (A) and gelatinised starches (B). Symbols: Maize 100 °C starch; Raw maize starch; Quinoa 100 °C starch; Raw quinoa starch. Selma-Gracia et al. [10].

However, although quinoa starch presented a greater susceptibility to digestion, gelatinised quinoa obtained about 25% less hydrolysis than gelatinised maize after 20 min, reaching higher total hydrolysis and, as a result, a slower hydrolysis rate.

4. Conclusions

The abovementioned results indicated that thermal parameters showed a reverse trend with hydrolysis, where quinoa displayed higher susceptibility to digestion than maize. The high hydrolysis and low slope kinetics from raw quinoa could suggest a potential starch for extending normoglycaemia in GSD patients. However, it would be necessary to evaluate the physiological consequences of this starch in an in vivo test.

Funding: This work was financially supported by grants QuiSalhis-Food (AGL2016-75687-C2-1-R) from the Ministry of Economy, Industry and Competitiveness (MEIC) and CYTED, la ValSe-Food (119RT0S67). The contract given to R. Selma-Gracia by LINCE (PROMETEO/2017/189) from the Generalitat Valenciana (Spain) is gratefully acknowledged.

References

1. Rosin, P.M.; Lajolo, F.M.; Menezes, E.W. Measurement and characterization of dietary starches. *J. Food Compos. Anal.* **2002**, *15*, 367–377, doi:10.1006/jfca.2002.1084.
2. Ahmadi-Abhari, S.; Woortman, A.J.J.; Oudhuis, A.A.C.M.; Hamer, R.J.; Loos, K. The influence of amylose-LPC complex formation on the susceptibility of wheat starch to amylase. *Carbohydr. Polym.* **2013**, *97*, 436–440, doi:10.1016/j.carbpol.2013.04.095.
3. Ratnayake, W.S.; Jackson, D.S. A new insight into the gelatinization process of native starches. *Carbohydr. Polym.* **2007**, *67*, 511–529, doi:10.1016/j.carbpol.2006.06.025.
4. Srichuwong, S.; Sunarti, T.C.; Mishima, T.; Isono, N.; Hisamatsu, M. Starches from different botanical sources I: Contribution of amylopectin fine structure to thermal properties and enzyme digestibility. *Carbohydr. Polym.* **2005**, *60*, 529–538, doi:10.1 016/j.carbpol.2005.03.004.
5. Ballester-Sánchez, J.; Gil, J.V.; Fernández-Espinar, M.T.; Haros, C.M. Quinoa wet-milling: Effect of steeping conditions on starch recovery and quality. *Food Hydrocoll.* **2019**, *89*, 837–843, doi:10.1016/j.foodhyd.2018.11.053.
6. Haros, M.; Blaszczak, W.; Perez, O.E.; Sadowska, J.; Rosell, C.M. Effect of ground corn steeping on starch properties. *Eur. Food Res. Technol.* **2006**, *222*, 194–200, doi:10.1007/s00217-005-0102-2.
7. Goñi, I.; Garcia-Alonso, A.; Saura-Calixto, F. A starch hydrolysis procedure to estimate glycemic index. *Nutr. Res.* **1997**, *17*, 427–437, doi:10.1016/s0271-5317(97)00010-9.
8. Sanz-Penella, J.M.; Laparra, J.M.; Haros, M. Impact of α-amylase during breadmaking on in vitro kinetics of starch hydrolysis and glycaemic index of enriched bread with bran. *Plant Foods Hum. Nutr.* **2014**, *69*, 216–221, doi:10.1007/s11130-014-0436-7.
9. Srichuwong, S.; Curti, D.; Austin, S.; King, R.; Lamothe, L.; Gloria-Hernandez, H. Physicochemical properties and starch digestibility of whole grain sorghums, millet, quinoa and amaranth flours, as affected by starch and non-starch constituents. *Food Chem.* **2017**, *233*, 1–10, doi:10.1016/j.foodchem.2017.04.019.
10. Selma-Gracia, R.; Laparra, J.M.; Haros, C.M. Potential beneficial effect of the hydrothermal treatment of starches from different sources on the in vitro digestion. *Food Hydrocol.* **2020**, *103*, 105687, doi:10.1016/j.foodhyd.2020.105687.

proceedings

MDPI

Proceedings

Variation of the Nutritional Composition of Quinoa According to the Processing Used †

Pablo Mezzatesta, Silvia Farah, Amanda Di Fabio and Raimondo Emilia *

Laboratorio de Investigación en Nutrición Aplicada, Facultad de Ciencias de la Nutrición Universidad Juan Agustín Maza, Lateral Sur del Acceso Este 2245 Dorrego Guaymallén Mendoza, CP 5519, Argentina; pablomezzatesta@gmail.com (P.M.); farahsilvia1@hotmail.com (S.F.); amandadifabio@gmail.com (A.D.F.)

* Correspondence: emilia.raimondo@gmail.com

† Presented at the 2nd International Conference of Ia ValSe-Food Network, Lisbon, Portugal, 21–22 October 2019.

Published: 4 August 2020

Abstract: Quinoa is consumed as a seed, flour, expanded, sprout (germinated) and activated (hydrated). The objective of this work was to determine the nutritional composition of the different preparations. The same batch of quinoa seeds was processed as flour, expanded, hydrated and germinated. It showed that there is a statistically significant difference of nutrients between all groups. For proteins, it varies from 12.78 ± 0.02 g/100 g in whole seed to 5.25 ± 0.01 g/100 g in the hydrated seed. In total fats, it varies from 7.80 ± 0.02 g/100 g in flour to 0.72 ± 0.01 g/100 g in sprouts. For fiber, the germinated quinoa provides the highest content (23.50 ± 0.01 g/100 g), whereas the hydrated quinoa the lowest content (8.71 ± 0.02 g/100 g). This shows how different preparations influence the nutritional contribution of quinoa. With this information, one can recommend different types of preparations depending on the type of nutrient that is wanted for consumption.

Keywords: expanded seed; flour; germinated seed; hydrated seed; quinoa

---

1. Introduction

According to Food and Agriculture Organization of the United Nations FAO [1], quinoa (*Chenopodium quinoa* Willd.) is a millenary crop that contributes to world food security. The protein content varies from 13 to 21% depending on the variety; it's essential amino acids profile is one of the most complete in the vegetable kingdom. That makes it an ideal food for populations with protein malnutrition [1]. Quinoa has an adequate content of dietary fiber, which decreases grain digestibility. The fiber contributes to granting satiety. It has been shown that the fatty acids of quinoa maintain quality due to the high natural value of vitamin E, which acts as a natural antioxidant [2]. Quinoa provides omega 6 polyunsaturated fatty acids (50% of their fat content) and omega 9 monounsaturated fatty acids (25% of their fat content) [3].

In Argentina, it is cultivated especially in the northwest; however, in the area of Mendoza, the crop shows good yields. Due to the type of harvest, it is ideal for small-scale production, benefiting small producers. In this context, Juan A. Maza University, the Provincial Legislature and the Family Agriculture Secretary of National Government are working on a project of the agronomist engineer Amanda Di Fabio [4].

In a global context with a strong demand for natural and nutritious products, quinoa is one of the Andean and ancestral crops most requested by consumers and with better economic prospects in recent years [5].

It is consumed as a seed, flour, expanded, sprout (germinated) and activated (hydrated). The popular belief is that all these forms have the same nutritional contribution. Taking into account this belief, the objective of this work was to determine the nutritional composition of the different forms of preparation for consumption of this grain.

2. Materials and Methods

This study was based on the same lot of quinoa, purchased from a local producer, which was washed, dried and free of saponins [6].

2.1. Ways of Preparation

Quinoa seed: It was washed and dried. This form of consumption is common.

Quinoa flour: The seed was grounded until a granulometry of 60 meshes (0.25 mm) was reached. In the present study, a laboratory mill was used and sieved prior to its analysis.

Activated quinoa (hydrated): Quinoa seeds were placed in water, at a room temperature of 20–25 °C, in a ratio of 1 part of seeds to 3 parts of water. They were left for 5 h. Then they were drained, cooked and used in different preparations. This method is common among vegans and vegetarians.

Quinoa was moistened overnight, drained, and placed in a glass jar upside down with a canvas as a lid for breathing. The next day it was moistened, drained and left face down again. The bottle was placed in the light and within 3 days the seeds began to sprout.

Expanded quinoa: A frying pan was heated over direct heat, with a small amount of oil. Quinoa seeds were added and after 5 min, they began to expand. The pan was removed from heat and the rest of the grains expanded.

2.2. Laboratory Analysis

To determine the nutritional composition of the different preparations of quinoa, the following laboratory determinations were made:

Humidity: Method of A.O.A.C 950.46. This was an indirect method that involved drying in an oven at 100–105 energy value (kcal) = (protein × 4) + (carbohydrates × 4) + (fat × 9). C, until constant weight was achieved.

Total fats: The direct method by extraction with ethyl ether (crude fat), Soxhlet gravimetric method (A.O.A.C. 960.39, 1990) was used.

Fibers: Dietary fiber (AOAC, 15th edition 1990) [7] was used.

Crude protein: The Kjeldahl method, (A.O.A.C. 928.08, 1990), determining nitrogen, using 6.25 as a protein conversion factor, was used.

Ashes: The direct Method (A.O.A.C. 923.03, 1990), involving incineration in a muffle (at 500 ± 10 Energy value (kcal) = (protein × 4) + (carbohydrates × 4) + (fat × 9). C), until constant ash weight was achieved, was used.

Carbohydrates: Carbohydrates were determined by difference, using the following formula:

$$100 - (\text{Weight in grams [protein + fat + water + ash + fibers]}), \text{ in 100 g of food.}$$

Energy value: Energy value was calculated according the following equation:

$$\text{Energy value (kcal)} = (\text{protein} \times 4) + (\text{carbohydrates} \times 4) + (\text{fat} \times 9).$$

The conversion is 2000 kcal = 8400 kJ.

2.3. Statistical Analysis

The same batch of quinoa was taken and processed (treatment) as flour, expanded, hydrated and sprout. The samples were analyzed in triplicate. For the statistical analysis, ANOVA was applied first and then a multiple comparison test was applied to discriminate between the means, and the honestly significant difference procedure (HSD) of Tukey was also applied.

3. Results

3.1. Nutrients Content for Each Type of Preparation

There were statistically significant differences (a < 0.05) between the nutrients of each type of preparation (seeds, sprouts, expanded, hydrated and flour).

A statistically significant difference was shown between all groups of nutrients. For proteins, it varied from 12.78 ± 0.02 g/100 g in whole seeds to 5.25 ± 0.01 g/100 g in the hydrated seed (Figure 1). In total fats, it varied from 7.80 ± 0.02 g/100 g in flour to 0.72 ± 0.01 g/100 g in sprout (Figure 2). For fiber, the germinated quinoa provided the highest content (23.50 ± 0.01 g/100 g) and the hydrated the lowest content (8.71 ± 0.02 g/100 g). The energy value was (kJ/100 g): whole seed 1299, flour 1430, germinated 291, hydrated 594 and expanded 1368 (See Table 1).

3.2. Statistical Analysis of the Data

The ANOVA statistical test was applied to analyze differences in means, and the HSD Tukey Multiple Ranges test was subsequently applied to analyze the difference between groups. The ANOVA result was statistically significant (p = 0.000) for all the variables analyzed (protein, carbohydrates, total fat, saturated fat, ash, fiber, moisture, sodium, and energy value).

The HSD Tukey Multiple Range test gave the following results:

Proteins: showed statistically significant differences between all levels, with the whole grain being the one with the highest content and the activated seed being the one with the lowest content.

Carbohydrates: showed statistically significant differences between all levels, with the expanded being the one with the highest content and the outbreak being the one with the lowest content.

Total fats: showed statistically significant differences between all levels, with the flour being the one with the highest content and the sprout with the lowest content.

Saturated fats: There were three homogeneous levels (sprout-activated-expanded) and this group showed statistically significant differences with flour and grain levels. The flour was the one with the highest content and (sprout-activated-expanded) the one with the lowest content.

Ash: There were two homogeneous levels (activated-expanded) and this group showed statistically significant differences with the flour, sprout and grain levels.

Table 1. Centesimal composition of quinoa, for different forms of preparation [a].

	Seed		Flour		Sprout		Hydrated		Expanded		ANOVA p
Carbohydrates	59.36	(0.05)	55.42	(0.01)	9.64 [c]	(0.04)	27.94	(0.03)	69.86 [b]	(0.02)	0.000
Saturated fats	0.26	(0.01)	0.86 [b]	(0.02)	0.08 [c]	(0.00)	0.11	(0.00)	0.06	(0.01)	0.000
Trans fat	n.d. [d]		n.d. [d]		n.d. [d]		n.d. [d]		n.d. [d]		0.000
Ashes	2.23	(0.02)	2.29 [b]	(0.01)	0.93 [c]	(0.01)	1.22	(0.01)	1.22	(0.01)	0.000
Humidity	10.64	(0.02)	10.63	(0.01)	59.16 [b]	(0.01)	55.91	(0.01)	7.00 [c]	(0.01)	0.000
Dietary fiber	12.68	(0.01)	11.71	(0.02)	23.50 [b]	(0.01)	8.71 [c]	(0.02)	12.63	(0.01)	0.000
Energy value kcal	309	(0.5)	341 [b]	(0.5)	69 [c]	(0.0)	141	(0.6)	326	(0.6)	0.000
Energy value kJ	1299	(0.5)	1430 [b]	(0.0)	291 [c]	(0.6)	594	(0.6)	1368	(0.6)	0.000

[a] Mean (SD); [b] Indicates the highest value; [c] Indicates the lowest value; [d] n.d., not detected.

The HSD Tukey Multiple Range test gave the following results:

Proteins: showed statistically significant differences between all levels, with the whole grain being the one with the highest content and the activated seed being the one with the lowest content.

Carbohydrates: showed statistically significant differences between all levels, with the expanded being the one with the highest content and the outbreak being the one with the lowest content.

Total fats: showed statistically significant differences between all levels, with the flour being the one with the highest content and the sprout with the lowest content.

Saturated fats: There were three homogeneous levels (sprout-activated-expanded) and this group showed statistically significant differences with flour and grain levels. The flour was the one with the highest content and (sprout-activated-expanded) the one with the lowest content.

Ash: There were two homogeneous levels (activated-expanded) and this group showed statistically significant differences with the flour, sprout and grain levels.

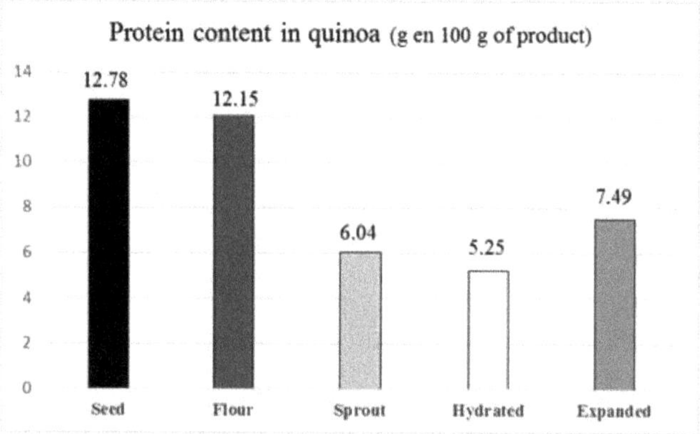

Figure 1. Content of proteins in quinoa for each of the treatments.

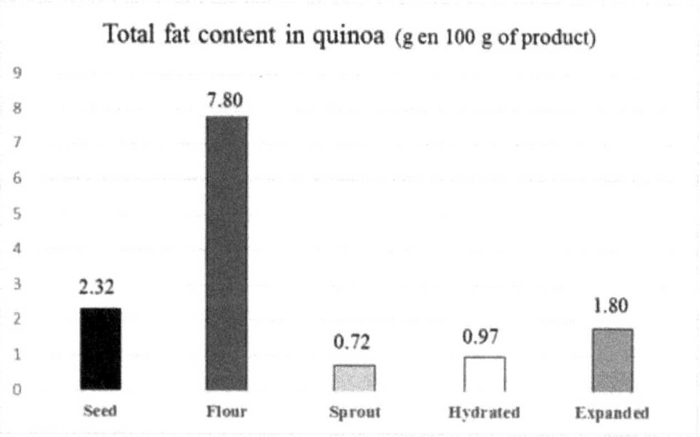

Figure 2. Content of total fat in quinoa for each of the treatments.

Humidity: There were two homogeneous levels (flour-grain) and this group showed statistically significant differences with the outbreak levels, activated and expanded. The outbreak was the one with the highest content and the expanded one with the least content.

Fiber: They showed statistically significant differences between all levels, the sprout being the one with the highest content and the activated seed the one with the lowest content.

Energy value: They showed statistically significant differences between all levels, the flour being the one with the highest content and the sprouts with the lowest content.

4. Discussion

The values found for the whole seeds are similar to those of the bibliography [8], except the fat content, which showed to be lower, since the grain was not ground for its determination. The determination of lipids was done in this way to simulate what would happen at the digestive level if the cuticles of the seed were not to be destroyed in the stomach. In the case of flour, the values are similar to those in the literature [1,8]. By making the lipid fraction more bioavailable, the energy contribution increases.

During the germination process, (quinoa sprout) nutrients diminished and the proportion of fiber increased, which made it ideal to grant satiety.

When the quinoa seed was placed in water (activated quinoa) it was hydrated by 205%, which meant that the content of all nutrients was reduced.

Finally, expanded quinoa increased its carbohydrate content, decreasing its protein and lipid value, and maintaining its fiber content.

5. Conclusions

This shows how different preparations influence the nutritional contribution of quinoa. With this information, it is possible to recommend different types of preparations depending on the type of nutrients that are relevant to add in a diet plan and wanted for consumption.

Funding: This research received no external funding.

Acknowledgments: This work was supported by Ia ValSe-Food-CYTED grant (119RT0567) and Juan Agustín Maza University.

References

1. FAO. 2011, La Quinua: Cultivo Milenario Para Contribuir a la Seguridad Alimentaria Mundial. Available online: http://www.fao.org/3/aq287s/aq287s.pdf (accessed on 8 May 2019).
2. Su-Chuen Ng Anderson, A.; Cokera, J.; Ondrusa, M. Characterization of lipid oxidation products in quinoa (*Chenopodium quinoa*). *Food Chem.* **2007**, *101*, 185–192, doi:10.1016/j.foodchem.2006.01.01.
3. Delatorre-Herrera, J.; Sánchez, M.; Delfino, I.; Oliva, M.I. La quinua (*Chenopodium quinoa* Willd.), un tesoro andino para el mundo. *Idesia (Arica)* **2013**, *31*, 111–114, doi:10.4067/S0718-34292013000200017.
4. Di Fabio, A. Proyecto de Producción, Comercialización y Promoción del Consumo de Quinoa en el Oasis Norte de Mendoza. 2018. Available online: https://www.legislaturamendoza.gov.ar/wp-content/uploads/2018/03/proyecto-de-quinoa-mdz-compressed.pdf (accessed on 8 May 2019).
5. INTA. 2012. Available online: http://intainforma.inta.gov.ar/?p=12134 (accessed on 8 May 2019).
6. Ahumada, A.; Ortega, A.; Chito, D.; Benítez, R. Saponinas de quinua (*Chenopodium quinoa* Willd.): Un subproducto con alto potencial biológico. *Rev. Colomb. Cienc. Quím. Farm.* **2016**, *45*, 438–469, doi:10.15446/rcciquifa.v45n3.62043.
7. AOAC Official Methods of Analysis. 1990. Available online: https://archive.org/stream/gov.law.aoac.methods.1.1990/aoac.methods.1.1990_djvu.txt (accessed on 24 April 2019).
8. INTA (Valerio Alejandro). Ciencia y tecnología de los cultivos industriales, Quinua. Año 3, N° 5—2013. ISSN 1853-7677. 2013. Available online: https://inta.gob.ar/sites/default/files/script-tmp-inta-revista-ciencia-y-tecnologa-de-los-cultivos-indu_4.pdf (accessed on 8 May 2019).

proceedings

MDPI

Proceedings

Making Nutritious Gluten-Free Foods from Quinoa Seeds and Its Flours †

Patricia Miranda Villa [1,2], Natalia Cervilla [1,3], Romina Mufari [1,4], Antonella Bergesse [1,2] and Edgardo Calandri [1,3,*]

1 Instituto de Ciencia y Tecnología de los Alimentos (ICTA)-FCEFyN-UNC, Córdoba 5000, Argentina; pmirandavilla@gmail.com (P.M.V.); lic.cervillanatalia@gmail.com (N.C.); romi_mufari@hotmail.com (R.M.); anto_bergesse@hotmail.com (A.B.)
2 Instituto de Ciencia y Tecnología de los Alimentos Córdoba (ICYTAC-CONICET), Córdoba 5000, Argentina
3 Escuela de Nutrición—Facultad de Ciencias Médicas-UNC, Córdoba 5000, Argentina
4 Instituto de Investigaciones Biológicas y Tecnológicas (IIBYT-CONICET), Córdoba 5000, Argentina
* Correspondence: edgardo.calandri@unc.edu.ar
† Presented at the 2nd International Conference of Ia ValSe-Food Network, Lisbon, Portugal, 21–22 October 2019.

Published: 5 August 2020

Abstract: Celiac disease is affecting around 1% of the world population and an effective treatment needs to exclude gluten. Quinoa is a high-quality gluten-free protein, and starch-rich endosperm, like the cereals. Protein contents and theoretical Protein Digestibility Corrected Amino Acid Score (PDCAAS) were evaluated in quinoas from Northwest and Centre of Argentina. A batter-type gluten-free quinoa bread was developed, showing good volume, taste, nutritional quality and a good long-lasting texture. Malted quinoa seeds' quality indicators rose until 48 h of germination; after that, an unpleasant taste was developed. Muffins made with that flour showed acceptable taste.

Keywords: amino acids; breads; germinates; muffins; proteins; quinoa

1. Introduction

Celiac disease is one of syndromes associated with irritable bowel and the most important of the enteropathies related to gluten [1]. Quinoa seeds possess a high-quality gluten-free protein and also have a starch rich endosperm, which makes this grain resembles cereals, and for this reason, it is usually called a pseudocereal. Quinoa protein has high levels of the essential amino acids and also is a good source of fiber, minerals and antioxidants [2]; but the lack of prolamins preclude quinoa protein from keeping the particular structure of bread crumb [3] and must be replace by starches and gums [4]. Therefore, the preparation of gluten-free breads requires the help of technology in food to satisfy the expectations of consumers.

The malting of seeds, which consists of allowing their germination in a controlled manner, can improve the absorption of nutrients because biochemical processes, which start when the seeds are moistened, release substances such as free amino acids, simple sugars and fatty acids, small molecules that easily overpass the intestinal epithelium.

The present work reassembles the studies accomplished by the group with quinoa seeds of the Centre and Northwest of Argentina and its employ in the development of batter type, gluten-free breads and muffins, with a focus on the nutritional quality of the products but also improving their taste and palatability.

2. Materials and Methods

2.1. Raw Materials

Northwest quinoa seeds were harvested from departments Molinos (2007–2008) and La Poma (2009–2011), Province of Salta, Argentina. Quinoas from the Centre region were originally Chilean "sea level" ecotype, varieties—Pichaman, Faro and Baer—cultivated in Río Cuarto, Argentina (2011). Bread preparation: quinoa, lupin and rice flours, hydroxypropyl methylcellulose (HPMC), sodium stearoyl lactylate (SSL), sugar and salt were mixed dry into a bowl. Fresh yeast was added, previously dispersed in warm water (28 °C), while mixing in a planetary mixer (280 rpm, 5 min). Water addition was completed in this step. The final mixture was loaded in a mold and leavened at 35 °C and relative humidity of 60% for 1 h. Finally, the dough was placed in the oven for 35 min at 180 °C. The bread loaf was left to cool at room temperature for 2 h before analysis.

2.2. Gravimetric Measures

The tests were carried out as described Cervilla et al. [5], Proximal Analysis, amino acid profile, Chemical Scoring (CS) and protein digestibility corrected amino acid score (PDCAAS), as described in Cervilla et al. [6]; the physical characteristics of products were determined as in Mufari et al. [7]; affective testing (consumer testing) and statistical analysis applied according to in Miranda-Villa et al. [8].

3. Results and Discussion

3.1. Physical Properties

Northern seeds were larger than those of the central region (Table 1). The weight of 1000 seeds (W1000) varied from 2.05 to 2.70 g for the central region, quite different to the northern ones, with values no lesser than 3.0 g for one thousand seeds, but the bulk density fluctuated between 0.55 and 0.73 g mL^{-1}, without a clear distinction between seed types. The size for northern seeds (equivalent diameter) was between 1.64 and 2.01 mm and can be classified as medium to large size, according to IBNORCA [5], while seeds for central region should be considered as just medium size. The apparent density values (Table 1) provide useful information for the analysis of heat transfer through grains, in quality control, and in the evaluation, calculation and design of transport systems, cleaning and classification. Real density results are important in the design of storage, packaging, dehydration and transportation systems.

Table 1. Gravimetric, dimensional properties and proximal composition of quinoa seeds [a].

Code [b]	1000 Seeds Weight	True Density	Bulk Density	d_1	d_2	e	Equivalent Diameter	Ash	Lipids	Carbohyd	Poteins
	(g)	(g/mL)	(g/mL)	(mm)	(mm)	(mm)	(mm)				
CENTRAL REGION SEEDS											
PCh	2.47 ± 0.01 b	1.01629272	0.554 ± 0.04 d	1.88 ± 0.15 a	1.91 ± 0.17 a	1.12 ± 0.17 a	1.59 ± 0.15 a	3.51 ± 0.21 c	6.54 ± 0.04 c	72.33 ± 0.51	17.18 ± 0.25 a
P_1Rc	2.13 ± 0.01 a	0.98586377	0.730 ± 0.01 a	1.74 ± 0.11 c	1.81 ± 0.12 b	1.00 ± 0.10 c	1.47 ± 0.10 b	4.31 ± 0.22 a	6.43 ± 0.04 c	70.46 ± 0.42	18.25 ± 0.14 a
P_2Rc	2.70 ± 0.02 a	1.21518085	0.619 ± 0.01 c	1.90 ± 0.11 a	1.93 ± 0.13 a	1.14 ± 0.15 a	1.61 ± 0.12 a	3.39 ± 0.04 c	6.49 ± 0.04 c	72.02 ± 0.58	17.64 ± 0.45 a
FCh	2.44 ± 0.01 b	1.07205692	0.687 ± 0.01 b	1.76 ± 0.17 c	1.83 ± 0.15 b	1.06 ± 0.14 b	1.51 ± 0.14 b	4.02 ± 0.10 b	6.16 ± 0.17 d	73.68 ± 0.52	15.68 ± 0.17 b
F_1Rc	2.16 ± 0.01 c	1.23720517	0.589 ± 0.02 d	1.80 ± 0.13 b	1.84 ± 0.14 b	1.00 ± 0.18 c	1.42 ± 0.13 b	4.43 ± 0.16 a	6.92 ± 0.17 b	70.28 ± 1.95	17.61 ± 1.49 a
F_2Rc	2.20 ± 0.01 c	1.06173958	0.744 ± 0.01 a	1.74 ± 0.13 c	1.78 ± 0.11 b	1.01 ± 0.11 c	1.46 ± 0.10 b	3.89 ± 0.03 b	8.19 ± 0.05 a	70.50 ± 0.29	16.95 ± 0.18 a
BCh	2.29 ± 0.01 c	0.98119775	0.577 ± 0.02 d	1.70 ± 0.12 c	1.78b ± 0.15 b	0.97 ± 0.15 c	1.43 ± 0.13 c	3.56 ± 0.15 c	7.05 ± 0.01 b	71.44 ± 0.61	17.41 ± 0.28 a
B_1Rc	2.14 ± 0.01 c	1.21661019	0.729 ± 0.01 a	1.72 ± 0.18 c	1.77 ± 0.12 b	0.98 ± 0.10 c	1.44 ± 0.11 c	3.97 ± 0.04 b	7.27 ± 0.00 b	71.40 ± 0.18	16.87 ± 0.13 a
B_2Rc	2.05 ± 0.01 c	1.21884051	0.715 ± 0.01 a	1.70 ± 0.11 c	1.75 ± 0.14 b	0.96 ± 0.11 c	1.42 ± 0.11 c	4.14 ± 0.07 b	5.83 ± 0.24 e	72.30 ± 1.23	18.25 ± 0.14 a
NORTHERN SEEDS [c]											
2007	3.2 ± 0.1 b	1.19 ± 0.10 a	0.69 ± 0.01 d	2.08 ± 0.10 a	2.12 ± 0.13 a	1.00 ± 0.10 a	1.64 ± 0.10 a	2.65 ± 0.58 b	9.03 ± 0.44 b	71.79 ± 1.57	16.53 ± 0.55 b
2008	3.0 ± 0.2 a	1.24 ± 0.11 a	0.72 ± 0.01 c	2.09 ± 0.10 a	2.12 ± 0.14 a	1.02 ± 0.10 a	1.66 ± 0.10 a	3.25 ± 0.10 b	9.06 ± 0.20 b	70.61 ± 0.73	17.08 ± 0.43 b
2009	4.7 ± 0.1 d	1.15 ± 0.04 a	0.68 ± 0.01 b	2.47 ± 0.11 c	2.41 ± 0.14 b	1.38 ± 0.12 c	2.01 ± 0.08 c	3.04 ± 0.06 a	8.8 ± 0.41 b	74.41 ± 1.03	13.75 ± 0.56 a
2010	3.4 ± 0.1 c	1.28 ± 0.03 a	0.66 ± 0.02 a	2.12 ± 0.10 a	2.22 ± 0.12 a	1.11 ± 0.07 b	1.73 ± 0.07 b	3.00 ± 0.08 a	6.05 ± 0.50 a	76.81 ± 0.97	14.14 ± 0.39 a
2011	3.5 ± 0.1 c	1.26 ± 0.01 a	0.66 ± 0.01 a	2.20 ± 0.09 b	2.19 ± 0.0 a	1.15 ± 0.11 b	1.77 ± 0.08 b	3.22 ± 0.10 b	6.52 ± 0.35 a	73.5 ± 0.78	16.76 ± 0.33 b

[a] Averages ± standard deviations are reported, in dry base. Numbers with different letters in the same column for same seed origin are significantly different. d1 = seed width, d2 = seed length, e = thickness of the seed. [b] P, P1 and P2 Pichanan varieties; F, F1 and F2, Faro varieties; B, B1 and B2, Baer varieties, Ch: harvested in Chile. Rc: harvested in Rio IV. [c] ordered by harvest year.

3.2. *Chemical Properties*

As shown in Table 1, ash content tends to be a little higher in seeds from Centre region and the opposite seems to be happening with lipids. Calcium and Magnesium being the main components of minerals present in quinoa while heavy metals like Pb and Cd are negligible [2], which mean that this grain can be considered as a valuable source of minerals, such as Ca, Mg, K, P, Fe, Cu and Zn.

Quinoa oil present good nutritional qualities [9] and can be considered as having potential for oil extraction. Carbohydrate contents shown in Table 1 are lower than those of common cereals but close enough to consider the quinoa grains as similar to cereal, i.e., pseudocereal [10]. The total protein varied from 13.75 to 18.25%, but quinoas from the Centre region showed rather higher value. Table 2 compares the amino acid contents for the five northern quinoas with bibliographic data for cereals and milk. The clear advantages of quinoa in histidine, methionine and lysine can be seen. In Table 3, the Chemical Scoring (CS) and Protein Digestibility Corrected Amino Acid Score (PDCAAS) of quinoa protein for preschools, schoolchildren and adults are presented. Data were calculated for 2009 and 2010 lots of northern quinoas, and in both cases the limiting amino acids were methionine and cysteine for all the age groups. Nevertheless, in seven of the amino acids of Table 3, the CS of the two quinoa batches covers 86% or more of the needs. The Protein Digestibility Corrected Amino Acid Score (PDCAAS), shown in Table 3, were estimated from a protein digestibility of 80% [11]. These results reflect the high disposability of quinoa amino acids and are similar to those from [12] in spite of these authors employing a higher factor for PDCAAS calculations.

3.3. *Products Development*

As can be seen in Table 3, quinoa is deficient in sulfur-rich amino acids, which makes it necessary to improve the nutritional quality of quinoa products by appealing to other grains, such as legumes. Accordingly, a batter-type gluten-free bread, with quinoa and lupin as the main ingredients, was developed, including 41% of quinoa, 29% of rice and 18% of lupin flours. This formulation has notable advantages when compared with commercial breads This formulation has notable advantages when compared with commercial breads, such as reducing carbohydrates and lipids and at the same time, increasing the polypeptides' content. The presence of sulfur-rich amino acids and also of histidine and serine are important. The final formulation was the result of many assays in which the effect of the main flours and minor components such as starches, gums, stabilizer, antioxidants and preservatives were statistically evaluated through experimental mixture designs. It was observed that water content and granulometry of the quinoa meal directly influenced the firmness and specific volume, while leavening and hydrocolloids other than HPMC did not have a significant effect on the quality of the loaves.

Leavening and hydrocolloids other than HPMC did not have a significant effect on the quality of the loaves. Between the emulsifiers, just SSL improved the texture, the specific volume and structure of the bread crumb. The incorporation of defatted flours of quinoa and lupine improved bread flavor, but it made the structure collapse [13]. The optimized formulation almost doubles the protein content of commercial quinoa breads and, at the same time, exhibits lower levels of lipids and carbohydrates, with higher content of essential amino acids. Despite being in the lowest proportion, lupin is the main protein source in the formula. The actual offer in gluten-free products is superabundant in carbohydrates and poor in proteins, as some studies show [14]; alternatives such as those described above tend to overcome the problem.

Table 2. Amino acid profiles of western quinoa seeds.

Amino Acid	2007	2008	2009	2010	2011	Quinoa [a]	Wheat [a]	Rice [a]	Oats [a]	Corn [a]	Barley [a]	Milk [a]
aspartic acid	1.07 ± 0.00	0.69 ± 0.04	1.09 ± 0.06	0.84 ± 0.04	1.14 ± 0.07	0.88	0.49	0.81	1.06	0.60	0.67	0.26
glutamic acid	1.87 ± 0.00	1.81 ± 0.04	1.90 ± 0.1	1.47 ± 0.04	2.24 ± 0.07	1.43	4.17	1.62	2.92	1.80	2.77	0.76
serine	0.39 ± 0.00	0.15 ± 0.00	0.55 ± 0.03	0.11 ± 0.03	0.57 ± 0.24	0.44	0.56	0.43	0.66	0.47	0.48	0.20
histidine	0.40 ± 0.00	0.29 ± 0.01	0.40 ± 0.02	0.90 ± 0.03	0.79 ± 0.53	0.29	0.25	0.2	0.29	0.26	0.25	0.09
glycine	0.69 ± 0.00	0.50 ± 0.01	0.78 ± 0.04	0.66 ± 0.02	0.59 ± 0.26	0.62	0.42	0.39	0.66	0.35	0.45	0.07
threonine	0.43 ± 0.00	0.19 ± 0.00	0.43 ± 0.02	0.35 ± 0.02	0.61 ± 0.27	0.42	0.32	0.31	0.46	0.34	0.39	0.15
arginine	1.04 ± 0.00	0.87 ± 0.00	1.15 ± 0.08	0.89 ± 0.04	1.34 ± 0.07	0.84	0.42	0.65	0.88	0.40	0.56	0.11
alanine	0.58 ± 0.00	0.65 ± 0.01	0.58 ± 0.03	0.48 ± 0.01	0.54 ± 0.44	0.56	0.37	0.47	0.63	0.72	0.46	0.12
proline	0.77 ± 0.00	0.55 ± 0.02	0.35 ± 0.06	0.09 ± 0.03	1.48 ± 0.06	0.37	1.39	0.37	0.72	0.85	1.28	0.31
tyrosine	0.34 ± 0.01	0.07 ± 0.01	0.32 ± 0.02	0.28 ± 0.01	0.26 ± 0.01	0.34	0.28	0.28	0.46	0.36	0.37	0.16
valine	0.62 ± 0.01	0.12 ± 0.01	0.71 ± 0.08	0.56 ± 0.07	0.50 ± 0.01	0.54	0.49	0.43	0.71	0.46	0.59	0.20
methionine	1.26 ± 0.01	0.78 ± 0.03	0.13 ± 0.01	0.11 ± 0.03	1.51 ± 0.34	0.24	0.17	0.18	0.23	0.18	0.20	0.09
cysteine	0.27 ± 0.01	1.64 ± 0.08	0.06 ± 0.00	0.04 ± 0.00	0.69 ± 0.01	---	0.30	0.08	0.37	0.15	0.27	0.03
isoleucine	0.53 ± 0.01	0.44 ± 0.01	0.53 ± 0.03	0.43 ± 0.01	0.64 ± 0.01	0.43	0.44	0.30	0.53	0.35	0.42	0.16
leucine	0.86 ± 0.01	0.66 ± 0.02	0.88 ± 0.05	0.71 ± 0.02	1.01 ± 0.02	0.72	0.84	0.65	1.01	1.19	0.78	0.33
phenylalanine	0.52 ± 0.01	0.31 ± 0.01	0.52 ± 0.03	0.43 ± 0.01	0.58 ± 0.02	0.49	0.58	0.41	0.70	0.46	0.60	0.19
lysine	0.63 ± 0.00	0.49 ± 0.01	0.60 ± 0.03	0.60 ± 0.02	0.78 ± 0.01	0.67	0.25	0.30	0.52	0.25	0.41	0.27

[a] FAO. United Nations Organization for Agriculture and Food. Food and nutrition collection (1970). Content in the amino acids of food and biological data on proteins. Rome Italy. Results expressed in g/100 g of seed four.

Table 3. Chemical Scoring (CS) [a] and protein digestibility corrected amino acid score (PDCAAS) of quinoa protein.

Amino Acids	Year 2010			Year 2009		
	Pre-School Chillaren	**School Children**	**Adults**	**Pre-School Children**	**School Children**	**Adults**
	2–5 Years Old	**10–12 Years Old**		**2–5 Years Old**	**0–12 Years Old**	
Histidine	400	400	475	136.8	136.8	162.5
Isoleucine	128.6	128.6	276.9	125	110.7	269.2
Leucine	90.9	136.4	315.8	87.9	131.8	305.3
Lysine	86.9	114.5	315	69	90.9	250
Methionine ± Cysteine [b]	50.8	57.7	74.7	59.2	67.3	87.1
Phenylalanine ± Tyrosine	95.4	273.2	316.3	88.9	254.5	294.7
Threonine	86.2	104.6	325.6	82.4	100	311.1
Tryptophan	s/d	s/d	s/d	427.3	522.2	940
Valine	134	187.6	360.8	217.1	304	584.6
Chemical Scoring	**50.8**	**57.7**	**74.7**	**67.3**	**59.2**	**87.1**
PDCAAS [c]	**40.6**	**46.2**	**59.8**	**47.4**	**53.8**	**69.7**

[a] Dimensionless number. [b] Limiting amino acid (LAA) [c] Calculated as 80% of theoretical digestibility.

Malting is the result of allowing a controlled seed germination, during this process many nutritious substances are released, particularly small sugars, amino acids and peptides [15], easily absorbed by the gut [10]. Muffins are batter-type formulations similar to the above-described bread [8], commonly consumed for breakfast and snacks. In muffins, the leavening stage is not needed and, after shaking, the batter is directly baked and then cooled at room temperature. Three formulations were assayed, with 30% quinoa (whole flour, and germinated whole flours for 24 and 72 h) and 70% rice flours and compared with a control muffin made just with rice flour. While the taste and flavor of rice muffin was preferred, that with quinoa flour germinated for 24 h presented a good taste and a texture barely less accepted than the control. Non-germinated quinoa flour formulation had negative comments regarding a bitter aftertaste, making it not eligible. The muffin formulation made with flour of quinoa germinated during 72 h presented a disgusting taste, frequently described as strong vegetal-like aftertaste, and it was also discarded. It is possible that the bitter taste in the first case was due to residual saponins present in seeds, but the unpleasant flavor perceived in the last one should be related to the germination process and the release of new substances. Untrained judges are often not familiar with gluten-free products, which normally provide great sensory changes, and explain their preferences for rice formulation. The slight disadvantage in flavor of the 24-h germinated formulation with respect to Control is compensated by the improvement observed at the nutritional level. As can be seen in Figure 1, quinoa protein seems to rise up until 24–48 h of germination, when it stabilizes, while lipids seem to keep increasing. Conversely, carbohydrates describe a negative tendency, as a consequence of the metabolic activity in seeds building, among others, proteins and lipids. Similar tendencies can be seen in Figure 2 for free essential amino acids; with the exception of valine, the others show a plateau at 24–48 h of germination. This behavior is also observed in the soluble matter (Figure 2), where the six substances seems to mimic their tendencies, giving flat curve segments between 24–48 h, all maxima. Many of them are small molecules which can go across the inner gut membrane with relative facility.

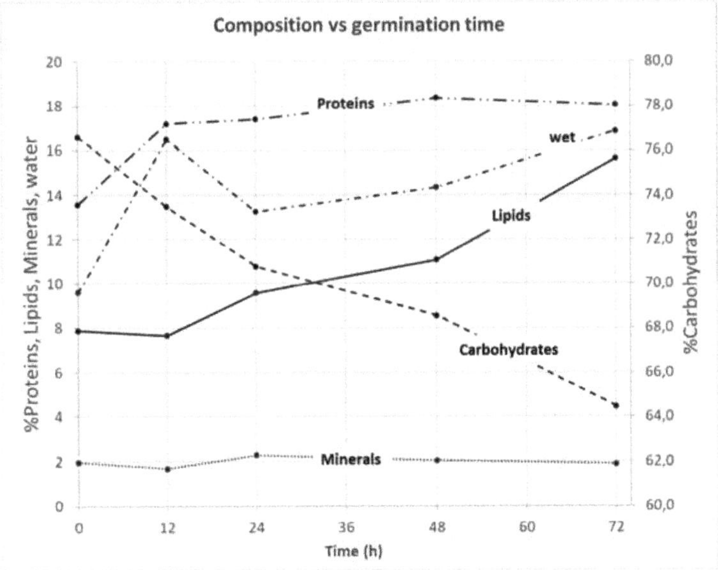

Figure 1. The change in proximate composition of quinoa flours along the germination time. All values expressed in dry base.

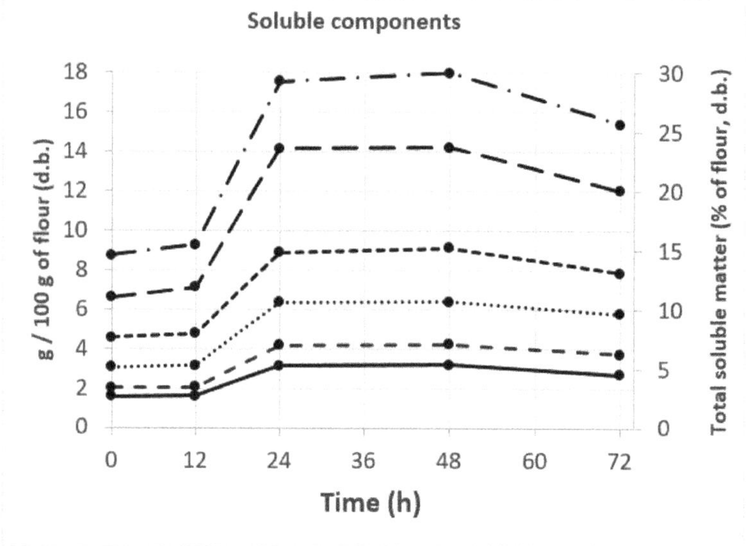

Figure 2. Changes in soluble components in quinoa flours, along the germination time.

4. Conclusions

In the present work, the nutritional quality of quinoa seeds and flours was shown. Quinoas from the Centre region where smaller than northern ones, with slightly higher protein content, but in all cases surpassing those of cereals. Lipid content was also greater than in cereals, and minerals ranged between 3% and 4%. Quinoa protein was rich in some essential amino acids but poor in methionine and cysteine, and mixing with sulfur-rich grain, such as legumes, is recommended. Gluten-free breads prepared with a mixture of quinoa and lupine flours showed good taste and texture and an improved nutritious quality. Muffins made with malted quinoa grain flour for 24 h showed a higher content of essential amino acids and soluble substances of small size, easily absorbed by the gut.

Funding: Authors wish to thank the Secretary of Science and Technology of the Universidad Nacional de Córdoba for the economic support.

Acknowledgments: This work was supported by grants Ia ValSe-Food-CYTED (119RT0567) and Secretaría de Ciencia y Técnica de la Universidad Nacional de Córdoba, Argentina.

References

1. Czaja-Bulsa, G. Non coeliac gluten sensitivity—A new disease with gluten intolerance. *Clin. Nutr.* **2015**, *34*, 189–194. doi:10.1016/j.clnu.2014.08.012.
2. Vidueiros, S.M.; Curti, R.N.; Dyner, L.M.; Binaghi, M.J.; Peterson, G.; Bertero, H.D.; Pallaro, A.N. Diversity and interrelationships in nutritional traits in cultivated quinoa (*Chenopodium quinoa* Willd.) from Northwest Argentina. *J. Cereal Sci.* **2015**, *62*, 87–93. doi:10.1016/j.jcs.2015.01.001.
3. Renzetti, S.; Rosell, C.M. Role of enzymes in improving the functionality of proteins in non-wheat dough systems. *J. Cereal Sci.* **2016**, *67*, 35–45. doi:10.1016/j.jcs.2015.09.008.
4. Ahmad, S.; Ahmad, M.; Rashid, H.; Ahmad, I. In fl uence of hydrocolloids on dough handling and technological properties of gluten-free breads. *Trends Food Sci. Technol.* **2016**, *51*, 49–57. doi:10.1016/j.tifs.2016.03.005.
5. Cervilla, N.; Mufari, J.; Calandri, E.; Guzmán, C. Propiedades Físicas de Semillas y Análisis proximal de harinas de *Chenopodium Quinoa* Willd Cosechadas en Distintos Años y Provenientes de la Provincia de Salta. *II Jorn. Investig. Ing. NEA Países Limítrofes* **2013**, *7*, 14–15.

6. Cervilla, N.; Romina, J.; Calandri, E.; Guzman, C. Determinación del contenido de aminoácidos en harinas de quinoa de origen argentino. Evaluación de su calidad proteica. *Actual. Nutr.* **2012**, *13*, 107–113. Available online: http://www.revistasan.org.ar/pdf_files/trabajos/vol_13/num_2/RSAN_13_2_107.pdf (accessed on 13 May 2017).
7. Mufari, J.; Miranda-Villa, P.; Bergesse, A.; Cervilla, N.; Calandri, E. Physico-chemical analysis and protein fraction compositions of different quinoa cultivars. *Acta Alim.* **2018**, *47*, 462–469. doi:10.1556/066.2018.47.4.9.
8. Miranda-Villa, P.P.; Mufari, J.R.; Bergesse, A.E.; Calandri, E.L. Effects of Whole and Malted Quinoa Flour Addition on Gluten-Free Muffins Quality. *J. Food Sci.* **2018**, *84*, 147–153. doi:10.1111/1750-3841.14413.
9. Mufari, J.R.; Gorostegui, H.A.; Miranda-Villa, P.P.; Bergesse, A.E.; Calandri, E.L. Oxidative Stability and Characterization of Quinoa Oil Extracted from Wholemeal and Germ Flours. *J. Am. Oil Chem. Soc.* **2020**, *97*, 57–66. doi:10.1002/aocs.12308.
10. García Luna, P.P.; López Gallardo, G. Evaluación de la absorción y metabolismo intestinal. *Nutr. Hosp.* **2007**, *22*, 5–13.
11. Tapia, M. Cultivos Andinos Sub-Explotados y su Aporte a la Alimentación. 2000. Available online: http://www.fao.org/tempref/GI/Reserved/FTP_FaoRlc/old/prior/segalim/prodalim/prodveg/cdrom/contenido/libro10/home10.htm (accessed on 24 March 2020).
12. Mota, C.; Santos, M.; Mauro, R.; Samman, N.; Matos, A.S.; Torres, D.; Castanheira, I. Protein content and amino acids profile of pseudocereals. *Food Chem.* **2016**, *193*, 55–61. doi:10.1016/j.foodchem.2014.11.043.
13. Miranda Villa, P.M. Efecto de la Adición de Harina de Quinoa y Lupino Dulce Sobre la Calidad Tecnológica de Panes Libres de Gluten. Ph.D. Thesis, FCEFyN–UNC, Córdoba, Argentina, 2019.
14. Naqash, F.; Gani, A.; Gani, A.; Masoodi, F.A. Gluten-free baking: Combating the challenges—A review. *Trends Food Sci. Technol.* **2017**, *66*, 98–107. doi:10.1016/j.tifs.2017.06.004.
15. Mäkinen, O.E.; Hager, A.S.; Arendt, E.K. Localisation and development of proteolytic activities in quinoa (*Chenopodium quinoa*) seeds during germination and early seedling growth. *J. Cereal Sci.* **2014**, *60*, 484–489. doi:10.1016/j.jcs.2014.08.009.

proceedings

MDPI

Proceedings

Influence of the Use of Hydrocolloids in the Development of Gluten-Free Breads from *Colocasia esculenta* Flour †

Jehannara Calle [1,2], Yaiza Benavent-Gil [2] and Cristina M. Rosell [2,*]

1 Food Research Institute for the Food Industry (IIIA), Carretera al Guatao km 3 ½, La Lisa, La Habana 17100, Cuba; yannaracalle@gmail.com

2 Institute of Agrochemistry and Food Technology (IATA-CSIC), C/ Agustin Escardino, 7, Paterna, 46980 Valencia, Spain; yaizabenavent@gmail.com

* Correspondence: crosell@iata.csic.es; Tel.: +34-9639-000-22; Fax: +34-9636-363-01

† Presented at the 2nd International Conference of Ia ValSe-Food Network, Lisbon, Portugal, 21–22 October 2019.

Published: 5 August 2020

Abstract: *Colocasia esculenta* represents an alternative non-gluten ingredient due to its healthy properties. The objective of this study was to explore the breadmaking potential of *Colocasia* spp. cormel flour combined with hydrocolloids (hydroxypropyl methylcellulose, xanthan gum, guar gum). A total of three formulations were tested. Breads were characterized by assessing their technological qualities: moisture, specific volume, volume, hardness and weight loss. The quality parameters were similar to other gluten-free breads. Overall, *Colocasia* spp. flour can be used to produce gluten-free breads with similar technological quality parameters than those previously reported with common gluten-free flours.

Keywords: bread; *Colocasia esculenta*; gluten-free; hydrocolloids

---

1. Introduction

In the field of the development of gluten-free bread, the replacement of the gluten network has been largely researched to meet the requirements of people with coeliac disease in order to improve the sensory, technological and nutritional quality of gluten-free breads [1]. Traditionally, most gluten-free products were made with rice, maize, sorghum, soy, buckwheat flours and starches from maize, potato, cassava, rice or beans [2]. Some researchers have used gums, hydrocolloids, enzymes and other ingredients to improve their minimal structure-building potential [3]. Thus far, hydrocolloids have been blended with GF (gluten-free) flours to improve the technological quality of end-products [4]. The *Colocasia esculenta* (L.) Schott (*Colocasia* spp.) rhizome is grown largely in Cuba [5]. Despite evidence of the positive benefits of this flour, there are few studies about its utilization in GF breadmaking [5]. The objective of this investigation was to explore the breadmaking potential of *Colocasia* spp. cormel flour combined with hydrocolloids (Hydroxypropylmethylcellulose: HPMC, xanthan gum, guar gum).

2. Materials and Methods

2.1. Materials

Cormels from freshly *Colocasia* spp. MC-2012, harvested at 9–13 months of maturity, were collected from the National Institute of Tropical Food Research Farms in Cuba, and the rhizomes were milled [5]. Hydroxypropylmethylcellulose (HPMC, Methocel™ K4M) was donated by Dow Pharma & Food Solutions (La Plaine Saint Denis, France). Guar gum–3500 and food-grade xanthan gum were acquired from EPSA (Valencia, Spain) and Jungbunzlauer (Wulzeshofen, Austria), respectively. All other ingredients were acquired in the local market and all reagents were of analytical grade.

2.2. Flour Characteristics and Baking Process

The methods used to determine the flour characteristics—moisture, total nitrogen, fat, ash and crude fiber content—were AACC, 1999 [6]; 78 AOAC, 1990; AACCI 79 Method 44-15.02; AACCI methods 46-12.01; AACC Method 30- 25.0; AACC Method 08-01.01 and AOAC Method 973.18, respectively. Carbohydrate content was estimated by difference. Water binding capacity (WBC) was analyzed according to the method described by [2]. Bread recipe was based on those previously selected [7,8] (Table 1) but replacing the flour with *Colocasia* spp. flour. Mixing was carried out in a Robot Coupe RM8 (Barcelona, Spain) at speed 3 for 8 min, and the rest of the breadmaking conditions were as previously reported, following the method described by [3].

Table 1. Gluten-free bread recipes.

Ingredients	F1	F2	F3
Flour (g)	100	100	100
Water (g)	227	227	227
Salt (g)	1.5	1.5	1.5
Compressed Yeast (g)	3	3	3
Sugar (g)	2	2	2
Oil (g)	2	2	2
HPMC (g)	-	2	0.29
Xanthan gum (g)	-	-	0.21
Guar gum (g)	-	-	0.50

100% of rhizome flour (F1); 100% of flour blended with hydrocolloids (F2 and F3).

2.3. Quality Assessment of the GF Breads

Quality parameters were evaluated as previously described by [9]. Bread moisture content and volume were evaluated following [6] and the rapeseed displacement method, respectively. Specific volume and weight loss during baking (cm^3/g) were calculated as the ratio between the volume of the bread and its weight and weighing the pans before and after baking, respectively. These measurements were replicated three times. The hardness was evaluated using a Texture Analyzer TA-XT2i (Stable Micro Systems, Surrey, UK).

2.4. Statistical Analysis

The data were reported as means ± standard deviation of two replicates. The effect of different additives was analyzed by ANOVA, using Statgraphics Centurion XVII software (Bitstream, Cambridge, MA, USA).

3. Results

3.1. Proximate Composition of Raw Flour

The proximate composition of the *Colocasia* spp. flour was as follows: moisture: 6.33% ± 0.02%, protein: 8.28% ± 0.07%, ash: 5.04% ± 0.00%, fat: 0.53% ± 0.00%, crude fiber: 4.4% ± 0.2% and carbohydrate: 75.4% ± 0.3%. Despite *Colocasia* spp. cormels exhibiting a high carbohydrate content, it contained an important amount of minerals and fiber. The protein content was lower than that reported for teff and buckwheat flours but higher than maize, rice, cassava or sweet potato flours [10–12].

3.2. *Effect of Additives in Colocasia spp. Cormel Flour-Based Breads*

The moisture content and weight loss were significantly different ($p < 0.05$) and as the moisture content decreased, the weight loss increased (Figure 1). In general, the addition of hydrocolloids resulted in higher moisture content than that observed in F1 breads. They also showed lower weight loss during baking, likely due to the water retention ability of hydrocolloids to retain water molecules. As expected, breads were visually dark, which could be attributed to the natural color of raw *Colocasia* spp. flour, which displayed 81.05 ± 0.36, 1.17 ± 0.07 and 14.63 ± 0.45 for L^*, a^* and b^*, respectively.

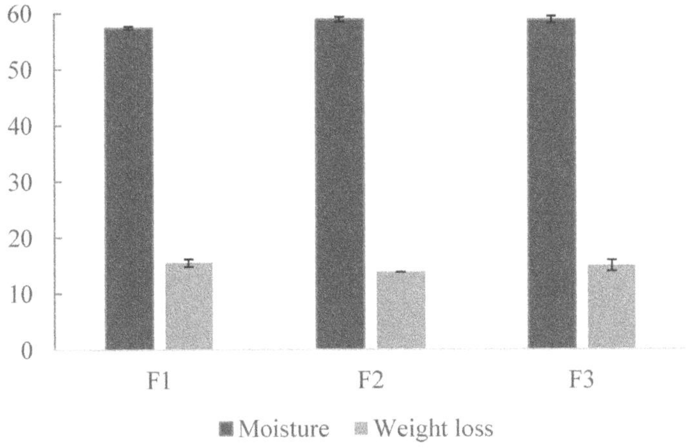

Figure 1. Weight loss and moisture content: 100% of rhizome flour (F1); 100% of flour blended with hydrocolloids (F2 and F3).

Hardness values ranged from 263 ± 38 to 361 ± 17 g. Overall, F1 breads displayed the highest values. In gluten-free breads, hydrocolloids usually led to lower crumb hardness [13]. Nonetheless, Sasaki [14] explained that their effect seems to be also dependent on the flour used; thus, it seems that, with cormel flour, the strengthening action of the hydrocolloids was not enough to hold the carbon dioxide during proofing and baking. In general, the bread without inclusion of hydrocolloids showed less hardness and a higher specific volume (1.74 ± 0.03 mL/g).

4. Conclusions

In order to develop a bread with a softer breadcrumb structure from *Colocasia* spp., flour is not required to include HPMC in the form of either a blend of HPMC, guar gum and xanthan gum at levels included. In this study, it can be concluded that hydrocolloids had a significant effect on the moisture content and weight loss of breads, but their effect was dependent on the type of hydrocolloids. Overall, *Colocasia* spp. flour can be used to produce GF breads with similar technological quality parameters to those previously reported with common GF flours.

Funding: Spanish Ministry of Science, Innovation and Universities (RTI2018-095919-B-C21), the European Regional Development Fund and Generalitat Valenciana (Project Prometeo 2017/189).

Acknowledgments: The authors want to thank Ia ValSe-Food-CYTED (119RT0567) for financing the participation in the meeting.

References

1. Capriles, V.D.; Arêas, J.A.G. Novel approaches in gluten-free breadmaking: Interface between food science, nutrition, and health. *Compr. Rev. Food Sci. Food Saf.* **2014**, *13*, 871–890, doi:10.1111/1541-4337.12091.
2. Cornejo, F.; Rosell, C.M. Influence of germination time of brown rice in relation to flour and gluten free bread quality. *J. Food Sci. Technol.* **2015**, *52*, 6591–6598, doi:10.1007/s13197-015-1720-8.
3. Calle, J.; Benavent-Gil, Y.; Rosell, C.M. Development of gluten free breads from Colocasia esculenta flour blended with hydrocolloids and enzymes. *Food Hydrocol.* **2020**, *98*, 105243, doi:10.1016/j.foodhyd.2019.105243.
4. Masure, H.G.; Fierens, E.; Delcour, J.A. Current and forward looking experimental approaches in gluten-free bread making research. .*J. Cereal Sci.* **2016**, *67*, 92–111, doi:10.1016/j.jcs.2015.09.009.
5. Calle, J.; Benavent-Gil, Y.; Garzón, R.; Rosell, C.M. Exploring the functionality of starches from corms and cormels of *Xanthosoma sagittifolium*. *Int. J. Food Sci. Technol.* **2019**, *54*, 2494–2501, doi:10.1111/ijfs.14207.
6. AACC. *Method 56-30.01 Water Hydration Capacity of Protein Materials*, 11th ed.; Method 44-15.02 Moisture—Air-Oven Methods, Approved Methods of Analysis; AACC International: St. Paul, MN, USA, 1999.
7. Calle, J.; Villavicencio MNd Rosell, C.M.; Bernabé-Marques, C.J. Influencia de mezclas de hidrocoloides en la reología de la masa del pan libre de gluten. *Cienc. Tecnol. Aliment.* **2014**, *2*, 37–42.
8. Marco, C.; Rosell, C.M. Breadmaking performance of protein enriched, gluten-free breads. *Eur. Food Res. Technol.* **2008**, *227*, 1205–1213, doi:10.1007/s00217-008-0838-6.
9. Matos, M.E.; Rosell, C.M. Quality Indicators of Rice-Based Gluten-Free Bread-Like Products: Relationships Between Dough Rheology and Quality Characteristics. *Food Bioprocess Technol.* **2013**, *6*, 2331–2341, doi:10.1007/s11947-012-0903-9.
10. Hager, A.-S.; Arendt, E.K. Influence of hydroxypropylmethylcellulose (HPMC), xanthan gum and their combination on loaf specific volume, crumb hardness and crumb grain characteristics of gluten-free breads based on rice, maize, teff and buckwheat. *Food Hydrocol.* **2013**, *32*, 195–203, doi:10.1016/j.foodhyd.2012.12.021
11. Pasqualone, A.; Caponio, F.; Summo, C.; Paradiso, V. M.; Bottega, G.; Pagani, M. A. Gluten-free bread making trials from cassava (manihot Esculenta Crantz) flour and sensory evaluation of the final product. *Int. J. Food Prop.* **2010**, 13, 562–573, doi:10.1080/10942910802713172.
12. Yadav, A. R.; Guha, M.; Tharanathan, R.; Ramteke, R. Changes in characteristics of sweet potato flour prepared by different drying techniques. *LWT Food Sci. Technol.*, **2006**, *39*, 20–26, doi:10.1016/j.lwt.2004.12.010.
13. Liu, X.; Mu, T.; Sun, H.; Zhang, M.; Chen, J; Fauconnier, M. L. Influence of different hydrocolloids on dough thermo-mechanical properties and in vitro starch digestibility of gluten-free steamed bread based on potato flour. *Food Chem.* **2018**, *239*, 1064–1074, doi:10.1016/j.foodchem.2017.07.047
14. Sasaki, T. Effects of xanthan and guar gums on starch digestibility and texture of rice flour blend bread. *Cereal Chem.* **2018,** *95*, 177–184, doi:10.1002/cche.10024.

proceedings

MDPI

Proceedings

Physicochemical and Techno-Functional Characterization of Native Corn Reintroduced in the Andean Zone of Jujuy, Argentina †

María Alejandra Giménez, Cristina Noemí Segundo, Manuel Oscar Lobo and Norma Cristina Sammán *

Facultad de Ingeniería, CIITED-CONICET—UNJu, Ítalo Palanca 10, San Salvador de Jujuy 4600, Argentina; malejandragimenez@gmail.com (M.A.G.); segundocristina@gmail.com (C.N.S.); mlobo58@gmail.com (M.O.L.)

* Correspondence: normasamman@gmail.com

† Presented at the 2nd International Conference of Ia ValSe-Food Network, Lisbon, Portugal, 21–22 October 2019.

Published: 5 August 2020

Abstract: The chemical and techno-functional properties of nine maize races from the Andean zone of Jujuy, Argentina, in the process of reintroduction, were determined. Principal component analysis (PCA) was applied to establish the differences between them. The breeds studied showed high variability in their chemical and techno-functional properties, which would indicate that their applications in the food industry will also be differentiated. The PCA analysis allowed us to group them into four groups, and the Capia Marron and Culli races showed unique properties, mainly in the formation of gels.

Keywords: Andean; native corn; physicochemical; race; techno-functional

1. Introduction

The native corn of the province of Jujuy, Argentina, has been the foundation of the alimentary culture of the Andean region for centuries. However, the way of feeding has changed with the progression into modernity, and this has led to the loss of numerous maize races, affecting the biodiversity of this crop. Currently, there are very few communities that are suppliers of corn seeds that were cultivated in the past [1]. However, the Puna, Quebrada and Valles regions are considered amongst the most important in situ Andean maize germplasm banks in the country.

The functional properties of corn flours indicate their possible uses in the food industry [2]. For example, the soft endosperm hydrates much better because starches are easily reached by water, as they have fewer bodies of zein surrounding the endosperm [3]. Likewise, the amylose–amylopectin ratio in starches explains the differences in the granular structure, the physico-chemical properties, the swelling power, the viscosity, and the gelatinization temperature, the firmness of the gel, the retrogradation and the susceptibility to enzymatic attack [4]. The incorporation of starches with high amylopectin content into flour has a beneficial effect in bakeries, as higher moisture delays retrogradation and extends shelf life [5]. The same does not occur in pasta, since it produces less firmness and greater stickiness [6]. There are also certain characteristics of the corn that make it suitable for producing specific foods, and for their use in beverages, as they are 85% grain and 15% cob [7].

The re-insertion of native breeds is of great importance for biodiversity, and the knowledge of their technological aptitudes is fundamental in determining their possible industrial applications. The objective of this work was to analyze the nutritional composition and the techno-functional

properties of the integral flours of nine Andean maize races, and to group them by applying principal components analysis (PCA).

2. Materials and Methods

2.1. Raw Materials, Sowing, Pollination and Harvest

The genetic material was provided by the INTA-Pergamino germplasm bank. The identification and origin of the genetic material is described in Table 1.

The sowing, pollination and harvesting was carried out in the experimental field of the Research Institute for Family Farming (IPAF INTA, Tilcara, Jujuy, Argentina) for two consecutive years (2017 and 2018). The ear corn were dried in the sun on metal sheets for 7 days (average temperatures: daytime 26 °C and nighttime 10 °C), then they were transported to the laboratory for analysis.

2.2. Chemical Composition, Amylose Content and Endosperm Hardness

The macronutrients of the different races were determined by Official Methods of Analysis [8]: humidity at 105 °C (AOAC 930.15), lipids by acid hydrolysis (AOAC 922.06), total nitrogen and proteins (AOAC 984.13), and ash by carbonization at 550 °C (AOAC 925.10). The amylose content was determined via the colorimetric method [9]. The hardness of the endosperm was determined by the hectoliter weight technique [3].

2.3. Properties of Hydration, Absorption of Oil and Thermal

The water holding capacity (WHC) at 30 °C and the oil holding capacity (OHC) at 70 °C were determined by the method of Ahmed et al. (2016). Both properties are expressed as g/g flour.

The thermal properties of the flours were analyzed by differential scanning calorimetry (DSC Q2000). The samples were prepared directly in capsules, with a final solids concentration of 25%, and they were heated from 20 to 130 °C at a rate of 10 °C/min in the presence of nitrogen. The initial temperature (To), peak temperature (Tp), final temperature (Tf) and enthalpy (ΔH/g) were determined from the area corresponding to each peak.

Table 1. Origin and identification of plant material.

Identifier	Race	Province	Department	Location
ARZM 09342	Capia blanco	Jujuy	Tumbaya	La Ciénaga
ARZM 09411	Capia Garrapata	Jujuy	Tilcara	Huacalera
ARZM 09417	Capia Marrón	Jujuy	Tilcara	Huacalera
ARZM 09332	Cristalino Amarillo	Jujuy	Tumbaya	Huachi Chocana
ARZM 09154	Culli	Jujuy	Tilcara	Potrerillos
ARZM 09435	Cuzco	Jujuy	Tumbaya	Huajra
ARZM 09144	Rojo	Jujuy	Tilcara	Pampa Grande
ARZM 09424	Morocho	Jujuy	Tilcara	Quebrada de la Huerta
ARZM 09418	Perlita	Jujuy	Tilcara	Huacalera

2.4. Firmness of Gels and Stability in Refrigeration

The gels were prepared from a flour/distilled water dispersion (3.5 ± 0.01 g flour in 25 ± 0.01 g water). The dispersion was heated to boiling with constant stirring for 3 min with a consistent heating plate temperature (temperature about 93 °C). They were then poured into cylindrical containers (3.5 m internal diameter, 4 cm high), allowed to stand for 25 min at room temperature and stored at 4 °C for 24 h for stabilization of the gel. The texture analysis was performed at room temperature on a texture analyzer TA.XT plus (Stable Micro Systems Ltd., Surrey, UK) equipped with Texture Exponent Lite software for Windows. A compression cycle was performed at a constant speed of 0.5 mms^{-1} to a depth of 8 mm of the sample, followed by a return to the original position. The force-time curve was obtained, and was used for the determination of firmness (the maximum force observed before the fracture). The stability of the gels was determined at 4 °C by measuring the syneresis for 96 h [10].

2.5. Statistical Analysis

All measurements were in triplicate. Data were analyzed using XLSTAT software (V2008.1.50162). Linear correlations between any two samples were estimated by Pearson correlation analysis. Principal component analysis (PCA) was carried out to reduce the dimensions of variables and to visualize the similarities among different samples.

3. Results and Discussion

3.1. Raw Materials

The nine races of corn planted were identified as Capia, Culli, Morocho, Garrapata, Perlita, Brown Capia, Red, Crystalline Yellow and Cuzco (Figure 1). For the identification of the different breeds, the manuals provided by the germplasm bank INTA-Pergamino were used.

3.2. Chemical Composition, Amylose Percentage and Endosperm Hardness

Table 2 presents the chemical composition and hardness of the endosperms of the different races. According to the endosperm hardness, the races were ordered from the hardest to the softest, as follows: Perlita, Morocho, Rojo, Cristalino amarillo, Culli, Capiamarrón, Capia, Garrapata and Cuzo. The moisture varied between 9.01% and 10%. The protein content varied between 7% and 12%. The Morocho and Capia Marron breeds had the highest and lowest protein contents, respectively. Culli and Perlita maize had the lowest (3.8%) and highest (5.7%) lipid content respectively. No correlation was observed between the hardness of the endosperm and the protein or ash content. However, a positive correlation was observed between lipid content and hectolitre weight. The amylose content varied between 15 and 29 mg/100 g. A tendency to increase the content of amylose was observed with the increase in hardness of the endosperm [11].

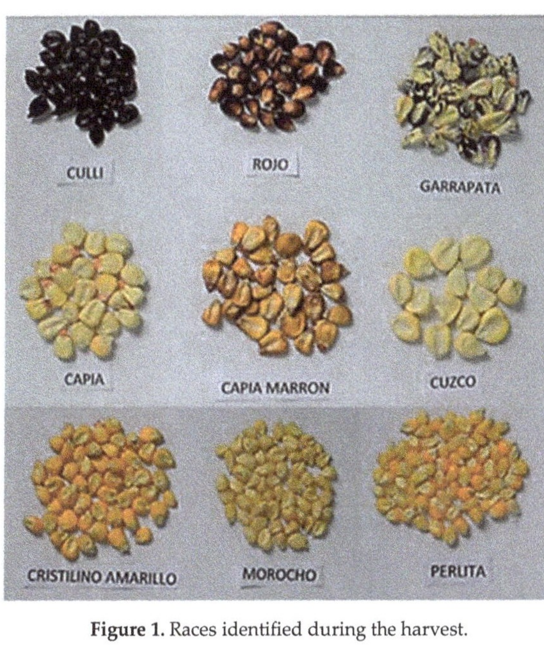

Figure 1. Races identified during the harvest.

3.3. Technological and Thermal Properties of Whole Meal Flours

The values of WHC were between 2.3 at 30 °C and 3.6 at 70 °C (Table 2). The Perlita race showed the lowest values, which could be attributed to the low fiber, starch and protein contents. WHC values, in general, increased at 70 °C due to the swelling of the starch associated with the gelatinization process. The Perlita race had the lowest OHC, which was significantly different from the rest of the samples, possibly due to the higher lipid content and lower protein content [12]. The peak gelatinization temperature for the different corn races varied between 67 and 69 °C. There were no significant differences in enthalpy of gelatinization between the starches of the different maize races, indicating that similar energies are required for gelatinization. The temperature range of gelatinization (ΔT) varied between 13 °C (Cuzco) and 25 °C (Culli), which suggests a wide diversity in gelatinization properties. The wholemeal flours are a very complex matrix; the presence of proteins and fibers hinders the access to water of the starch, increasing the Tf and ΔT. The gels presented a wide variation in firmness (16–81 Kp), and the lowest firmness corresponds to the gels of Cuzco (very soft endosperm) and the greatest to the Perlita (hard endosperm). However, no correlation was observed with endosperm hardness. The stability of the gels over 96 h in refrigeration was represented by syneresis between 4% and 32%; the greater stability was found for the Culli, and the least for Capia Marron. The percentage of syneresis correlated positively with the amylose content.

3.4. Principal Component Analysis (PCA)

Figure 2a shows the PCA applied to the chemical composition and techno-functional properties. The F1 factor explains 28.3% of the variability of the chemical composition, attributed mainly to the contents of ash and lipids, while F2 (21.09%) is attributed to the amylose content. The sample with the highest positive F1 score was the Culli race, and the highest F2 was Perlita. The main techno-functional properties that contributed to F1 were WHC and firmness. In F2, the variability was contributed by the OHC and syneresis. Only the variable ΔH is close to the crossing of the axes, which suggests its low contributions to the variability of the properties. It corresponds for F1 to the race Culli, and for F2 to Perlita. In Figure 2b, two groups of races with similar characteristics are distinguished: one composed of the Capia, Cuzco and Rojo races, with high endosperm hardness and protein content; and the other composed of the Perlita, Cristalino Amarillo and Morocho races, which formed firm gels. The Capia Marron and Culli breeds showed distance from the other two groups, suggesting that their properties are unique.

3.5. Principal Component Analysis (PCA)

Figure 2a shows the PCA applied to the chemical composition and techno-functional properties. The F1 factor explains 28.3% of the variability of the chemical compositions, attributed mainly to the contents of ash and lipids, while F2 (21.09%) is attributed to the amylose content. The samples with the highest positive F1 score correspond to the Culli race, and the highest F2 to Perlita. The main techno-functional properties that contribute to F1 are WHC and firmness. In F2, the variability was contributed by the OHC and syneresis. Only the variable ΔH is close to the crossing of the axes, which suggests its minimal contribution to the variability of the properties. It corresponds for F1 to the race Culli, and for F2 to Perlita. In Figure 2b, two groups of races with similar characteristics are distinguished: one composed of the Capia, Cuzco and Rojo races, with high endosperm hardness and protein content; and the other composed of the Perlita, Cristalino Amarillo and Morocho races, which formed firm gels. The Capia Marron and Culli breeds showed distance from the other two groups, suggesting that their properties are unique.

4. Conclusions

A great variability in the chemical composition and technical-functional properties has been observed among the nine integral flours of the Andean corn races. According to their broad techno-functional behaviors, they could be used in different technological processes to produce integral foods. The analysis of the main components revealed that Culli and Capia Marron differ from the other races, mainly due to the formation of gels of intermediate hardness, and the high stability of Culli. The recovery of these races through the revaluation of their technological properties will contribute to the maintenance of biodiversity and the food security of rural families in the Andean region of Jujuy, Argentina.

Acknowledgments: This work was supported by grant Ia ValSe-Food-CYTED (Ref. 119RT0567) and Consejo Nacional de Investigaciones Científicas y Técnicas (CONICET) and Secretaría de Ciencia y Técnica y Estudios Regionales (SECTER), Universidad Nacional de Jujuy (Argentina).

References

1. Ramos, R.S.; Hilgert, N.I.; Lambaré, D.A. Agricultura Tradicional y riqueza de maíces (*Zea mays*). Estudio de Caso en Caspalá, provincia de Jujuy, Argentina. *Bol. Soc. Argent. Bot.* **2013**, *48*, 607–621.
2. Hung, P.V.; Maeda, T.; Morita, N. Waxy and high-amylose wheat starches and flours—Characteristics, functionality and application. *Trends Food Sci. Technol.* **2006**, *17*, 448–456, doi:10.1016/j.tifs.2005.12.006.
3. Salinas-Moreno, Y.; Aguilar-Modesto, L. Effect of maize (*Zea mays* L.) grain hardness on yield and quality of tortilla. *Ing. Agric. Y Biosist.* **2010**, *2*, 5–11, doi:10.5154/r.inagbi.2010.08.009.
4. Wolf, B. Polysaccharide functionally through extrusion processing. *Curr. Opin. Colloid Interface Sci.* **2010**, *15*, 50–54, doi:10.1016/j.cocis.2009.11.011.
5. Morita, N.; Maeda, T.; Miyazaki, M.; Yamamori, M.; Miura, H.; Ohtsuka, I. Effect of substitution of waxy-wheat flour for common flour on dough and baking properties. *Food Sci. Technol. Res.* **2002**, *8*, 119–124, doi:10.3136/fstr.8.119.
6. Gimenez, A.; Drago, S.; De Gree, F.D.; Gonzalez, R.; Lobo, M.; Samman, N. Rheological, functional and nutritional properties of wheat/broad bean (*Vicia faba*) flour blend for pasta formulation. *Food Chem.* **2012**, *134*, 200–206, doi:10.1016/j.foodchem.2012.02.093.
7. Saldaña, E.; Rios-Mera, J.; Arteaga, H.; Saldaña, J.; Samán, C.M.; Selani, M.M.; Villanueva, N.D.M. How does starch affect the sensory characteristics of mazamorra morada? A study with a dessert widely consumed by Peruvians. *Int. J. Gastron. Food Sci.* **2018**, *12*, 22–30, doi:10.1016/j.ijgfs.2018.01.002.
8. AOAC. *Official Methods of Analysis*, 16th ed.; Association of Official Analytical Chemists: Washington, DC, USA, 1995.
9. Juliano, B.; Perez, C.; Blakeney, A.; Castillot, D.; Kongseree, N., Laingnelet, B.; Lapis, E.; Murty, V.; Webb, B. International cooperative testing on the amylose content of milled rice. *Starch/Staerke.* **1981**, *33*, 157–162, doi:10.1002/star.19810330504.

10. Eliasson, A.C.; Kim, H.R.; Changes in rheological properties of hydroxypropyl potato starch pastes during freeze-thaw treatments. I. A rheological approach for evaluation of freeze-thaw stability. *J. Texture Stud.* **1992**, *23*, 279–295, doi:10.1111/j.1745-4603.1992.tb00526.x.
11. Robutti, J.L.; Borras, F.S.; González, R.J.; Torres, R.L.; De Greef, D.M. Endosperm properties and extrusion cooking behavior of maize cultivars. *Food Sci. Technol.* **2002**, *35*, 663–669, doi:10.1006/fstl.2002.0926.
12. Ahmed, J.; Al-Attar, H.; Arfat, Y.A. Effect of particle size on compositional, functional, pasting and rheological properties of commercial water chestnut flour. *Food Hydrocoll.* **2016**, *52*, 888–895, doi:10.1016/j.foodhyd.2015.08.028.

proceedings

MDPI

Proceedings

Enrichment of Protein and Antioxidants of Cupcake with Moringa (*Moringa oleifera*) Leaf Powder and Sensorial Acceptability †

Alejandra Chinchilla, Susana Rubio-Arraez, Marisa L. Castelló and María Dolores Ortolá *

Institute of Food Process Engineering for Development, Universitat Politècnica de València, Camino de Vera s/n, 46022 Valencia, Spain; alechinchita2193@gmail.com (A.C.); suruar@upvnet.upv.es (S.R.-A.); mcasgo@upv.es (M.L.C.)

* Correspondence: mdortola@tal.upv.es

† Presented at the 2nd International Conference of Ia ValSe-Food Network, Lisbon, Portugal, 21–22 October 2019.

Published: 5 August 2020

Abstract: *Moringa oleifera* plants have an extensive range of bioactive compounds (carbohydrates, phenolic compounds, lipids and fatty acids, proteins and functional peptides). These molecules may be included in several food matrices, such as bakery products, to improve their nutritional values. For that, the aim of this study was to replace the part of wheat flour with 1%, 2.5%, 5% and 10% of moringa leaf powder in cupcakes, assessing their antioxidant capacity, protein content and sensorial acceptability. The results showed that proteins and antioxidant capacity directly increased with moringa content. However, according to the tasters, these moringa-rich cupcakes were too dark.

Keywords: antioxidant capacity; cupcakes; moringa; proteins; sensorial analysis

---

1. Introduction

Moringa is a plant grown in the north of India, Spain, Africa, the Middle East and South America, being *Moringa oleifera* the most cultivated species. Their leaves contain several nutrients such as vitamins, minerals, amino acids, β-carotenes, antioxidants, fiber and proteins (19–29%) with low caloric levels [1–3]. Due to their nutritional components, dried leaves have been used to fortify different food products such as soups, pasta, breads, cakes and cookies [4].

In this regard, the aim of this study was to evaluate the level of wheat flour replacement by dried moringa powder (0%, 1%, 2.5%, 5% and 10%) on the water, protein and antioxidant capacity in cupcakes. Furthermore, a sensorial assessment was carried out to estimate the acceptability of these products.

2. Materials and Methods

2.1. Materials

Dried moringa leaves, grown in the Valencia region, were ground in a blender (Thermomix, TM31, Vorwerk, Wupertal, Germany) for 3 min at maximum speed (10,000 rpm). Then, the powder was passed through a 0.1 mm mesh sieve and it was stored at room temperature in a sealed glass jar protected from light.

2.2. Preparation of Cupcakes

Cupcakes were prepared with the following components (*w/w*): 25% of eggs, 25% of sucrose, 25% of wheat flour and/or leaf moringa powder, 12% of sunflower oil, 12% of milk and 2% of baking powder. Firstly, eggs were beaten in an electrical blender (Kenwood Ltd, KM240 serie, New Lane, Havant, UK) for ten minutes at maximum rate. After that, sucrose was added, and the mixture was blended for 5 more minutes. Then, milk and oil were incorporated, and the batter was mixed for 2 min at a low rate. Finally, wheat flour/moringa and baking powder were aggregated in the batter, and it was blended for 5 min at a low rate. The batter was left to stand in a refrigerator for 30 min. Next, muffin paper cups (60 × 35 mm) were filled with 65 g of batter and they were baked at 145 °C for 23 min in a semi industrial oven (Rational AGD-86899, Landsberg am Lech, Germany).

Depending on the degree of wheat flour replacement, five formulations were prepared: M0% (without moringa), M1%, M2.5%, M5% and M10% when the degree of replacement was 1%, 2.5%, 5% and 10%, respectively.

2.3. Analytical Determinations

Moisture content was obtained by means of the gravimetric method [5]. Protein content was determined by the Kjeldahl method [6]. An adaptation of the spectrophotometric DPPH method [7,8] was used to analyze the antioxidant capacity. For that, the percentage of free radical DPPH inhibition was registered according to the following equation:

$$\%\ DPPH\ _{reduction} = 100 \cdot \left[\frac{A_{control} - A_{sample}}{A_{control}}\right]$$

where: $A_{control}$ = absorbance of initial *DPPH* (without sample) and A_{sample} = absorbance after 30 min of sample addition [9].

2.4. Sensorial Analysis

Acceptance of cupcakes formulated with different percentages of moringa powder (M1%, M2.5% and M5%) was analyzed with a panel composed of 30 panellists between 18 and 60 years of age. This test was performed in a sensorial room according to the rule ISO 4121:2003 [10]. On one hand, a hedonic scale was considered to find out the scores that the taster gave to the formulations depending on the attributes analyzed. On the other hand, a just about right (JAR) scale was used in order to know if the intensity of the attribute should be higher or lower.

3. Results and Discussion

Percentages of antioxidant capacity, water and proteins of the cupcakes are shown in Table 1. As can be seen in the table, moringa powder contains a high antioxidant capacity, which comes from compounds such as vitamin C, E and β-carotene. The antioxidant capacity of cupcakes increased linearly with respect to the percentage of moringa used (% DPPH inhibition = 7.4568 + 7.1074% Moringa, $R^2 = 0.9301$), showing that the antioxidant power of moringa powder persists after baking.

Table 1. Percentages of water, protein and DPPH inhibition in cupcakes (M) depending on the amount of dried leaf moringa powder (0%, 1%, 2.5%, 5% and 10%) and in moringa powder.

Product	Antioxidant Capacity (% Inhibition)	% Water	% Protein
M0%	1.18 ± 0.06 [a]	26.5 ± 0.8 [a]	5.81 ± 0.12 [a]
M1%	11.41 ± 0.07 [b]	28.2 ± 0.4 [ab]	6.08 ± 0.02 [b]
M2.5%	38.81 ± 0.01 [c]	26.8 ± 1.5 [a]	6.52 ± 0.09 [c]
M5%	40.86 ± 0.01 [d]	27.8 ± 1.0 [ab]	7.18 ± 0.10 [d]
M10%	76.5 ± 0.2 [e]	29.33 ± 0.07 [b]	8.07 ± 0.10 [e]
Moringa powder	80.16 ± 0.07 [f]	7.3 ± 0.3	29.6 ± 0.3

Equal letters mean homogeneous groups obtained in the ANOVA analysis.

The persistence of the antioxidant activity of moringa after baking may be related to the interaction between the components of the cupcake (mainly proteins) and the active compounds (phenols) [11]. As was expected, protein content increased with the increase in moringa percentage in the product following a linear fitting (% Protein = 5.8946 + 0.2264% Moringa, R^2 = 0.9864) in coherence with the results found by other authors [12]. Bear in mind that in each formulation, for the theoretical protein content of the traditional components [13] and the protein obtained in the moringa powder (Table 1), no loses were found after baking in these conditions.

The relationship between water and protein content may be related to the strong ability of the dried leaf moringa powder to bind to water, due to its high amount of proteins [14,15]. The moringa powder protein has several polar aminoacids such as serine, treonine, proline and glutamine [1]. These polar aminoacids interact with water molecules, so cupcakes that contain moringa powder showed a higher water retention ability.

The lower scores obtained in the attributes of flavor, aroma, aspect and color when the percentage of moringa increased were probably due to the lower height reached in these cases and the great intensity of the green color, although the mechanical properties were well assessed without significant differences, irrespective of the level of moringa in the formulation (Figure 1).

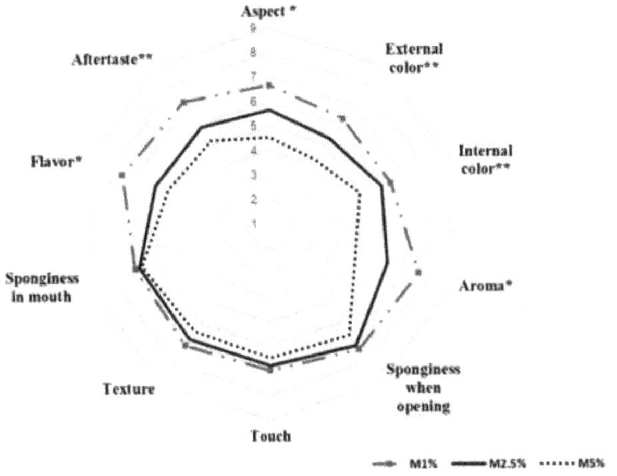

Figure 1. Hedonic scores of cupcakes formulated with moringa. (* significant level > 95%, significant level > 99%).

According to the results shown in Figure 2, the greater darkening of cupcakes with the moringa determined the purchase intention.

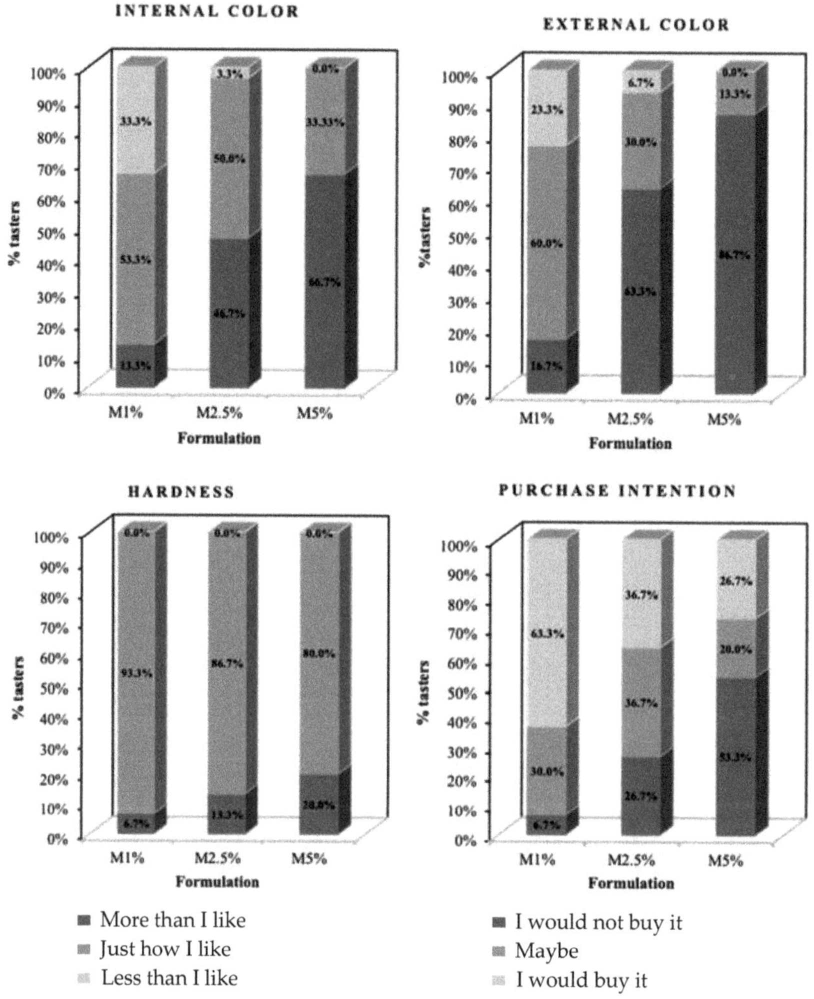

Figure 2. Just about right assessment (JAR) of cupcakes formulated with moringa.

4. Conclusions

It is feasible to formulate cupcakes with moringa leaf powder, increasing their antioxidant capacity and protein content. However, more studies should be carried out in order to improve their aspect, flavor and aroma since tasters penalized these attributes as the moringa concentration increased.

Acknowledgments: This work was supported by grant Ia ValSe-Food-CYTED (Ref. 119RT0567) and by the project "New Crops Addressing Climate Change: Moringa and Stevia" (Ref. AGCOOP_A/2018/026 AVFGA—Generalitat Valenciana).

References

1. Moyo, B.; Masika, P.; Hugo, A.; Muchenje, V. Nutritional characterization of Moringa (*Moringa oleifera* Lam.) leaves. *Afr. J. Biotechnol.* **2011**, *10*, 12925–12933, doi:10.5897/AJB10.1599.
2. Srinivasamurthy, S.; Yadav, U.; Sahay, S.; Singh, A. Development of muffin by incorporation of dried Moringa oleifera (Drumstick) leaf powder with enhanced micronutrient content. *Int. J. Food Sci. Nutr.* **2017**, *2*, 173–178.

3. Abbas, R.K.; Elsharbasy, F.S.; Fadlelmula, A.A. Nutritional Values of Moringa oleifera, Total Protein, Amino Acid, Vitamins, Minerals, Carbohydrates, Total Fat and Crude Fiber, under the Semi-Arid Conditions of Sudan. *J. Microb. Biochem. Technol.* **2018**, *10*, 56–58, doi:10.4172/1948-5948.1000396.
4. Oyeyinka, A.; Oyeyinka, S. Moringa oleifera as a food fortificant: Recent trends and prospects. *J. Saudi Soc. Agric. Sci.* **2018**, *17*, 127–136, doi:10.1016/j.jssas.2016.02.002.
5. AOAC. *Official Methods of Analysis of AOAC International*; Determination of moisture content; The Association of Official Analytical Chemists: Gaithersburg, MD, USA, 2000.
6. AOAC. *Official Methods of Analysis of AOAC International. Method 928.08*; The Association of Official Analytical Chemists: Gaithersburg, MD, USA, 1974.
7. Brand-Williams, W.; Cuvelier, M.; Berset, C. Use of a free radical method to evaluate antioxidant activity. *LWT Food Sci. Technol.* **1995**, *28*, 25–30, doi:10.1016/S0023-6438(95)80008-5.
8. Turkemen, N.; Sari, F.; Veliouglu, Y.S. The effect of cooking methods on total phenolics and antioxidant activity of selected green vegetables. *Food Chem.* **2004**, *93*, 713–718, doi:10.1016/j.foodchem.2004.12.038.
9. Turkemen, N.; Sari, F.; Veliouglu, Y.S. The effect of cooking methods on total phenolics and antioxidant activity of selected green vegetables. *Food Chem.* **2004**, *93*, 713–718, doi:10.1016/j.foodchem.2004.12.038
10. ISO 4121 SA. *Guidelines for the Use of Quantitative Responses Scales*; International Organization for Standardization: Geneva, Switzerland, 2003.
11. Bourekoua, H.; Różyło, R.; Gawlik-Dziki, U.; Benatallah, L.; Nasreddine Zidoune, M.; Dziki, D. Evaluation of physical, sensorial, and antioxidant properties of gluten-free bread enriched with *Moringa Oleifera* leaf powder. *Eur. Food Res. Technol.* **2017**, *244*, 189–195, doi:10.100700217-017-2942-y.
12. Abdull, R.; Ahmad, F.; Ibrahim, M.D.; Kntayya, S.B. Health benefits of Moringa oleifera. *Asian Pac. J. Cancer Prev.* **2014**, *15*, 8571–8576, doi:10.7314/apjcp.2014.15.20.8571.
13. Bedca. 2019. Spanish Food Composition Database. Available online: http://www.bedca.net/bdpub/index.php (accessed on 1 June 2019).
14. Sun-Young, K.; Chang-Ho, C. Quality Characteristics of Noodles added with Moringa oleifera Leaf Powder. *J. East Asian Soc. Diet. Life* **2017**, *27*, 321–331, doi:10.17495/easdl.2017.6.27.3.321.
15. Aryee, A.; Boye, J. Comparative study of the effects of processing on the nutritional, physicochemical and functional properties of lentil. *J. Food Process. Pres.* **2017**, *41*, 12824, doi:10.1111/jfpp.12824.

proceedings

MDPI

Proceedings

Plinia peruviana "Yvapurũ" Fruits and Marmalade from Paraguay: Autochthon Products with Antioxidant Potential †

Lourdes Wiszovaty, Silvia Caballero, Cristian Oviedo, Fernanda Ozuna and Laura Mereles *

Universidad Nacional de Asunción, Facultad de Ciencias Químicas, Dirección de Investigaciones; Departamento Bioquímica de alimentos, P.O. 1055, San Lorenzo, Paraguay; lourdesw@qui.una.py (L.W.); scaballero@qui.una.py (S.C.); cristian_ovi_ro@hotmail.com (C.O.); andita21ozuna@hotmail.com (F.O.)

* Correspondence: laurameleres@qui.una.py

† Presented at the 2nd International Conference of Ia ValSe-Food Network, Lisbon, Portugal, 21–22 October 2019.

Published: 6 August 2020

Abstract: In this study, we aimed to describe the antioxidant content and physicochemical characteristics of the marmalade and fruits of *Plinia peruviana* "Yvapurũ", harvested in Paraguay. The morphological characteristics, moisture, pH, vitamin C, total phenols, and anthocyanins were analyzed in mature wild and cultivated fruit samples. The values of anthocyanins (282–288 mg of cyanidin 3-O-glucoside/100 g) and total phenols in fruits and marmalades (214–224 and 719–817 mg GAE/100 g, respectively) make this fruit and its marmalade byproduct potential sources of phenolic compounds of interest to the dye, flavoring, and antioxidant industries.

Keywords: *Plinia peruviana;* antioxidants; total phenols; marmalade; anthocyanins; Yvapurũ

---

1. Introduction

The systematic investigation of native fruits and their derived products promotes their re-evaluation and contributes to a better exploration of the national species, which motivates new economic activities [1]. The biodiversity of fruits in Paraguay is considerable; however, the limited information on their composition results in a lack of use on the agronomic, industrial, and medicinal levels. *Plinia peruviana* (Poir.) Govaerts "Yvapurũ" is a perennial tree of the Myrtaceae family that is distributed in the Southern Cone of America in the regions of Central Department, Cordillera, Ñeembucu, and Paraguarí in Paraguay. This species is currently accepted as *Eugenia guapurium* DC., *Guapurium peruvianum* Poir., *Myrciaria guapurium* (DC.) O. Berg, *Myrciaria cauliflora* auct. non (DC.) O. Berg, *Eugenia cauliflora* Miq., hom. illeg., *Plinia trunciflora* (O. Berg) Kausel, *Myrciaria trunciflora* O. Berg and *Myrciaria peruviana* (Poir.) Mattos var. trunciflora synonyms [2]. In the region, the chemical composition and biological activities of these fruits have been characterized and studied, highlighting a high total phenol content in the fruit with antioxidant activity. The content of phenols varies with the environment, plant growth, genetic variety, and the stage of maturation of the fruit factors, among others [2–4]. One of the most traditional products that uses all the parts of the fruit is marmalade. We aimed to describe the antioxidant content and physicochemical characteristics of *Plinia peruviana* "Yvapurũ" fruits harvested in Paraguay, as well as determine the physicochemical characteristics and content of total phenols in marmalade (Figures 1 and 2).

Figure 1. Fruit of *Plinea peruviana*.

Figure 2. Batches of Yvapurũ marmalade.

2. Materials and Methods

2.1. Sampling

Samples of wild and cultivated fruits were collected in the ripe state from Emboscada, Cordillera, Paraguay (Cabaña ITAPÉ) (Sample 1) and Caacupé, Coordillera, Paraguay (Sample 2) from the Paraguayan Institute of Agrarian Technology (IPTA), respectively. The handmade whole-fruit marmalades were elaborated from the Sample 1 collection by traditional washing, slicing, cooking, and sieving without additives in three different batches.

2.2. Processing of Samples

For the determination of the morphological and physicochemical characteristics and vitamin C content in the fruit, whole fruits were used. To analyze total anthocyanins and total phenols, the peel and seeds of the pulp were separated. All determinations were made in triplicate; the data obtained are expressed as the mean ± standard deviation.

2.3. Analytical Methods

For the morphological studies, 30 fruits were taken for each sample, their weight was measured in analytical balance (AYD, model HR 120, Bradford, England), longitudinal and transversal diameter (measured in cm), and pH was measured with a potentiometer (Accurate pH 900, Horiba, Kyoto, Japan) at 25 °C. For humidity and vitamin C analysis, official A.O.A.C. techniques were used [5]. All reagents used were analytical grade. All determinations were made in triplicate. The determination of total anthocyanins by the differential pH method [1] was based on monomeric anthocyanin color loss at pH 4.5 and the presence of color at pH 1. The absorbance was measured at 510 and 700 nm.

Total phenols were determined using the Folin–Ciocalteau method with some modifications based on a colorimetric oxide reduction reaction. The extraction was carried out as described by Rufino et al. [6].

2.4. Statistical Analysis

The data were recorded and processed in a form of the GraphPad Prism 5.0 program (GraphPad Software Inc., San Diego, CA, USA). To determine significant differences, $p \leq 0.05$ was considered.

3. Results and Discussion

3.1. Yvapurũ Fruit Characteristics

No significant differences were observed in the morphological characteristics of weight and longitudinal and transversal diameter (Table 1) between the fruit samples. This result showed that the analyzed fruits harvested in Paraguay are smaller than those of synonymous species fruits in the state of Paraná, Brazil. These fruits weigh between 6.4 and 11.4 g and their diameter is near to 2.65 cm [7,8].

Table 1. Physicochemical characteristics of *Plinia peruviana* "Yvapurũ" fruits.

Physical Characteristics	Sample 1	Sample 2
Longitudinal diameter (cm)	1.98 ± 0.19^{a}	2.07 ± 0.18^{a}
Transversal diameter (cm)	2.01 ± 0.18^{a}	2.12 ± 0.19^{a}
Weight (g)	5.5 ± 1.3^{a}	5.6 ± 1.4^{a}
pH	3.17 ± 0.04^{a}	3.13 ± 0.01^{a}
Moisture (g/100 g)	79.7 ± 0.5^{a}	78.8 ± 0.1^{b}
Vitamin C (mg/100 g fw)	9.87 ± 0.30^{a}	7.03 ± 0.02^{b}

Results are expressed as mean ± SD of three independent assays. Values in the same row with the same superscript letter are not significantly different ($p > 0.05$) as measured by Student's *t*-test.

The pH values in the fruits are acidic and differ from the results reported by other authors (pH = 3.6–4.3) for the synonymous species of *Plinia peruviana* harvested in Brazil [7] *P. cauliflora* and *P. trunciflora*. These characteristics are explained by soil type and different environmental conditions, as well as genetic variations in the varieties distributed at different latitudes [9]. Thus, this indicated that the characteristics and composition of native fruits vary with soil type, genetics, and environmental conditions during the development of the species [7].

The moisture values in whole fruits were lower than 80%, with a significant difference between the average and the lower percentage observed by other authors (85.9%) in Brazil [6]. These values, however, are similar to the values observed by Seraglio et al. [10] for whole fruits of "Jaboticaba" (79.63%). Significant differences were observed in the vitamin C content between the fruits (Table 1). The values were lower than those reported for whole fruit in Brazil for *Myrciaria cauliflora* (238 mg/ 100 g) by Rufino et al. [6].

3.2. Anthocyanins and Total Phenol Content in Fruits

Significant differences in anthocyanins and total phenol content in different parts of the fruit were observed. Total phenols were highest in the peel (Table 2). Anthocyanins in the Sample 1 collection peel were smaller than in the Sample 2 collection. It was reported that this fruit has up to 58.1 mg/100 g of anthocyanins in fresh whole fruit [6] and 298.8–426.3 mg/100 g in peel, which agrees with the values observed in the present work. In the pulp, they reported lower values, 0.071–2.024 mg/100 g [8], than what was observed in pulps and seeds, which are used for the preparation of handmade marmalade in Paraguay. The values obtained in the pulp and seed were higher than those reported by other authors for *Plinia cauliflora* pulp in Brazil (32.4 mg GAE/100 g), and lower than the values reported in fruit peel harvested in Minas Gerais, Brazil [8].

Table 2. Anthocyanins and total phenols in fruits of *Plinia peruviana* "Yvapurũ".

Parameter	Sample 1		Sample 2	
	Peel	Pulp + Seeds	Peel	Pulp + Seeds
Monomeric anthocyanins (mg/100 g of cyanide nidin 3-glucoside)	282 ± 3 [a]	18.6 ±1.1 [b]	288 ± 2 [c]	10.0± 0.8 [d]
Total phenols (mg GAE/100 g FW)	811 ± 21 [a]	719 ± 15 [b]	817 ± 31 [a]	749 ± 22 [b]

Results are expressed as mean ± SD of three independent assays. Values in the same row with the same superscript letter are not significantly different ($p > 0.05$) as measured by Student's *t*-test.

3.3. *Yvapurũ* Plinia peruviana *Marmalade Results*

Significant differences in the titratable acidity were observed between batches 2 and 3 (ANOVA and Tukey's post-hoc test, $p \leq 0.05$) as shown in Table 3.

Table 3. Characteristics of the fruit marmalade of *Plinia peruviana* "Yvapurũ".

Parameter	Batch 1	Batch 2	Batch 3
Soluble solids (°Brix)	68.7 [a]	68.9 [a]	66.2 [b]
pH	2.80 ± 0.06 [a]	2.82 ± 0.01 [a]	2.82 ± 0.06 [a]
Total solids (g/100 g)	3.21 ±0,30 [a]	3.02 ± 0.28 [a]	3.63 ± 0.88 [a]
Titratable acidity (g citric acid/100 g)	0.88 ± 0.02 [a,b]	0.82 ± 0.02 [a]	0.97 ± 0.06 [b]
Total phenols (mg GAE/100 g FW)	224 ± 11 [a]	223 ± 7 [a]	214 ± 10 [a]

Results are expressed as mean ± SD of three independent assays. Values in the same row with the same superscript letter are not significantly different ($p > 0.05$) as measured by ANOVA and Tukey's post-hoc test, $p \leq 0.05$.

The total phenol content did not show statistically significant differences (ANOVA and Tukey's post-hoc test, $p \leq 0.05$). The results showed that the "Yvapurũ" marmalade, with its acidic pH, its organic acid content (titratable acidity 0.828–0.937 g citric acid/100 g fw), and sugar content (66.2–68.9° Brix), allows for the natural conservation of the product without the addition of artificial additives.

The marmalade has about 27–30% fewer total phenols than the pulp + seed and peel fruits, and provides about 22 mg of the total phenols per 10 g serving (one tablespoon of marmalade). Major polyphenols described for these species are quercetin, gallic acid, cyanidin-3-O-glucoside, isoquercetin, 3,4-dihydroxybenzoic acid, and kaempferol. These polyphenols give the product its antioxidant potential and other bioactive properties, such as anti-inflammatory, antibacterial, antifungal, antiproliferative, antimutagenic, hypoglycemic, and hypolipidemic activities [9,10]. The pomace, obtained as a byproduct (skin and seeds), can be of use in the food industry.

4. Conclusions

The fruits of *Plinea peruviana* "Yvapurũ" are important sources of vitamin C, anthocyanins, and phenolic compounds. Anthocyanins are found mainly in the peel; however, phenols are distributed in the peel, pulp, and seeds. Phenolic compounds may be of interest to the food industry, as colorants, antioxidants, and flavorings. The marmalade provides polyphenols that give it added value. Studies on native fruits and their elaborated products should be furthered to characterize their unique chemical properties, with a possible denomination of their origin.

Acknowledgments: This work was supported by a grant from the Ia ValSe-Food CYTED Project (119RT0567). The authors are especially grateful to Cabaña Itapé and IPTA for the provision of the samples and the Facultad de Ciencias Químicas of the Universidad Nacional de Asunción for providing their facilities.

References

1. IICA. *Protocolos estandarizados para la valoración de frutos nativos del procisur frente a la creciente demanda por ingredientes y aditivos especializados (carotenoides, antocianinas y polifenoles)*; PROCISUR: Montevideo, Uruguay, 2018.
2. Instituto de Botánica Darwinion. Catálogo de Plantas Vasculares: *Plinia peruviana* (Poir) Govaerts. Available online: http://www.darwin.edu.ar/Proyectos/FloraArgentina/DetalleEspecie.asp?forma=&variedad=&subespecie=&especie=peruviana&genero=Plinia&espcod=21619 (accessed on 28 July 2020).
3. Bailão, E.F.L.C.; Devilla, I.A.; da Conceição, E.C.; Borges, L. Bioactive compounds found in Brazilian cerrado fruits. *Int. J. Mol. Sci.* **2015**, *16*, 23760–23783, doi:10.3390/ijms161023760.
4. Gurak, P.D.; De Bona, G.S.; Tessaro, I.C.; Marczak, L.D. Jaboticaba pomace powder obtained as a co-product of juice extraction: A comparative study of powder obtained from peel and whole fruit. *Food Res. Int.* **2014**, *62*, 786–792, doi:10.1016/j.foodres.2014.04.042.
5. Horwitz, W. (Ed.) *Official Methods of Analysis of AOAC International*, 17th ed.; AOAC International: Gaithersburg, MD, USA, 2000.
6. Rufino, M.S.; Alves, R.; Brito, E.S.; Pérez-Jiménez, J.; Saura-Calixto, F.; Mancini-Filho, J. Bioactive compounds and antioxidant capacities of 18 non-traditional tropical fruits from Brazil. *Food Chem.* **2010**, *121*, 996–1002, doi:0.1016/j.foodchem.2010.01.037.
7. Citadin, I.; Vicari, I.; da Silva, T.; Danner, M. Qualidade de frutos de jabuticabeira (*Myrciaria cauliflora*) sob influencia de duas condicioes de cultivo: sombramento natural e pleno sol. *R Bras. Agrociência* **2005**, *11*, 373–375.
8. Wagner Júnior, A.; Paladini, M.V.; Danner, M.A.; Moura, G.C.; Guollo, K.; Nunes, I.B. Aspects of the sensorial quality and nutraceuticals of Plinia cauliflora & fruits. *Acta Sci. Agron.* **2017**, *39*, 475. doi:10.4025/actasciagron.v39i4.35420.
9. Borges, L.; Conceição, E.C.; Silveira, D. Active compounds and medicinal properties of *Myrciaria* genus. *Food Chem.* **2014**, *153*, 224–233, doi:10.1016/j.foodchem.2013.12.064.
10. Seraglio, S.K.T.; Schulz, M.; Nehring, P.; Della Betta, F.; Valese, A.C.; Daguer, H.; Costa, A.C.O. Nutritional and bioactive potential of Myrtaceae fruits during ripening. *Food Chem.* **2018**, *239*, 649–656, doi:10.1016/j.foodchem.2017.06.118.

proceedings

MDPI

Proceedings

Sicana odorifera "Kurugua" from Paraguay, Composition and Antioxidant Potential of Interest for the Food Industry †

Coronel Eva, Caballero Silvia, Baez Rocio, Villalba Rocio and Mereles Laura *

Departamento de Bioquímica de Alimentos, Facultad de Ciencias Químicas, Universidad Nacional de Asunción, P.O. Box 1055, San Lorenzo, Paraguay; ecoronel@qui.una.com (C.E.); scaballero@qui.una.com (C.S.); rocio.baez15@gmail.com (B.R.); rvillalba@qui.una.com (V.R.)

* Correspondence: lauramereles@qui.una.py

† Presented at the 2nd International Conference of Ia ValSe-Food Network, Lisbon, Portugal, 21–22 October 2019.

Published: 6 August 2020

Summary: The aim of this study was to evaluate the physicochemical characteristics, centesimal composition and antioxidants of the *Sicana odorifera* pulp and the antioxidant potential of the seeds and fruit peel harvested in a culture of the city of San Lorenzo, Paraguay. These fruits harvested in Paraguay present an antioxidant potential, interesting for food industry, especially in a ripe and semi-ripe state, where the highest content of vitamin C and total phenols was observed, as well as the total antioxidant capacity (ABTS).

Keywords: *Sicana odorifera*; composition; antioxidants; Curcubitaceae; total phenols; vitamin C

---

1. Introduction

Worldwide, the search for food components as additives with functional properties represents a great demand, where biodiversity and the systematic study of species of interest is fundamental to generating knowledge about potential uses of native fruits, with known medicinal applications described in ethnobotany [1]. This specie is cultivated in several countries in South America and Central America, known as melao de croa, cassabanana, or red melon [2]. In Brazil, the composition of the fruit of the kurugua was recently described [2], however, the phytochemical composition of the fruit pulp has been little explored in Paraguay, despite being described as an autochthonous fruit, which limits its use at an industrial level, mainly due to the ignorance of its nutritional potential and consequently of its potential applications. The fruit of the kurugua is used in juices, jams and preserves, it is used as an infusion in popular wisdom for liver diseases [3], and some species of the same family Cucurbitaceae have known hepatoprotective properties [4]. It is known that antioxidants can contribute to liver protection, and they vary in their concentration and quality with the stage of maturity. The objective of the present work was to evaluate the physicochemical characteristics, centesimal composition and antioxidants of the *Sicana odorifera* pulp and to evaluate the antioxidant potential in the seeds and fruit peel harvested in a culture in San Lorenzo city, Paraguay.

2. Materials and Methods

2.1. Sampling

Samples of *Sicana odorifera* fruits "kurugua" were collected from healthy mature trees (12 years old) from an orchard placed at San Lorenzo, Central Department, Paraguay (GPS −25.3266340 N,

−57.4832020 E). The yearly average sunlight in the cultivation place ranges between 6 and 14 h per day. The orchard is used as a seedbed of the "Kurugua Poty" foundation, which ensured the traceability of the variety analyzed. Samples were collected for herbarium material for botanical identification by the Botánica Department of the Facultad de Ciencias Químicas of the Universidad Nacional de Asunción. Random sampling from the crop of fruits in three stages of maturity—unripe, semi-ripe, and ripe—was carried out. Fourteen kilograms of fruits were obtained. After the morphological measurements, the pulp was separated from the peel and seeds. The centesimal composition was made in the edible fraction (pulp) of ripe fruit.

2.2. *Physicochemical Characters*

Morphological studies were performed on unripe, semi-ripe and ripe fruits without previous treatment, as described by Mereles and Ferro [5]. The pH, titratable acidity and soluble solids were measured according to AOAC Methods [6]. A potentiometer (Accurate pH 900, Horiba, Kyoto, Japan) at 25 °C and an analytical balance (AYD HR 120, Bradford, England) were used. All measurements were made in triplicate. All the reagents used were of analytical grade.

2.3. *Chemical Analysis*

Moisture, protein, total carbohydrates, dietary fiber, total mineral content or ash were determined, all according to official AOAC methodologies [6]. To estimate the antioxidant potential of the fruit, the content of total phenols, monomeric anthocyanins, vitamin C and the total antioxidant capacity were evaluated by the ABTS method. The determination was made using the spectrofluorometric method 967.22 of AOAC [6]. For the measurements, an L-ascorbic acid calibration curve (2.5–20 μg/mL) was used. The results were expressed in mg of vitamin C per 100 g of pulp fresh weight. The determination of anthocyanins was carried out by the spectrophotometric method of differential pH, based on the color loss of the monomeric anthocyanins at pH 4.5 and presence of color at pH 1, measuring at 510 and 700 nm [1]. The final concentration of anthocyanins (mg/100 g) was calculated based on the volume of extract and sample fresh weight. It is expressed in cyanidinidine 3-glucoside (MW: 449.2 and ε: 26,900). An extraction of 2 g of lyophilized pulp with methanol:water (60:40) in an ultrasonic bath for 15 min was performed and subsequent centrifugation (15 min, 10,000× *g* rpm, 4 °C). After separating the supernatant, a second extraction was carried out with acetone:water (70:30) with the same treatment described. Extractions and measurements were carried out by triplicates. Total phenolic compounds (TPC) were determined by the Folin–Ciocalteau described by Singleton & Rossi [7], colorimetric method using a calibration curve obtained with gallic acid (0–120 μg/mL aqueous solution). The mixture was stirred and kept for 30 min at room temperature in the dark. The absorbance was measured spectrophotometrically at 765 nm against a blank reagent. For measurements, a gallic acid standard curve (concentration interval 0–120 μg/mL aqueous solution) was plotted at 765 nm in the UV-1800 spectrophotometer (Shimadzu, Kyoto, Japan). The results were expressed as mg of gallic acid equivalents (GAE) per 100 g of pulp (mg of GAE/100 g). The antioxidant activity was determined by the TEAC assay using the radical cation $ABTS^{+\bullet}$ [8] (Re et al. 1999). The $ABTS^{+\bullet}$ stock solution (7 mM) was prepared using ammonium persulphate $(NH_4)_2S_2O_8$ as the oxidant agent. The working solution of $ABTS^{+\bullet}$ was obtained by diluting the stock solution in ethanol to give an absorption of 0.70 ± 0.02 at λ = 734 nm. For measurements, a calibration curve with Trolox (0–500 μM aqueous solution) was plotted at 730 nm. The results were expressed as micromoles of Trolox equivalents (TEAC) per gram of pulp fresh weight.

2.4. *Statistical Analysis*

The results were expressed as means ± standard deviation (SD) from three independent replicates. Data were stored in a spreadsheet, when appropriate, to compare stages of maturity values (unripe, semiripe and ripe) ANOVA and Tukey's post-test were used. Values with $p \leq 0.05$ were considered as statistically significant with the assistance of Graph Pad Prism 5.0 software (GraphPad Software, Inc., San Diego, CA, USA) for the calculations.

3. Results and Discussion

3.1. Physicochemical Characters

The physicochemical characteristics of the whole fruit of *S. odorifera* were determined in three stages of maturity (Table 1). Variations in weight, diameter and height were observed among unripe fruits compared to semi-ripe and ripe fruits. These results are consistent with those reported by Paula Filho et al. (2015) in *S. odorifera* harvested in Brazil, for the weight and average diameter of the ripe fruits (2510 g, 9.72 cm, respectively) [3]; however these authors reported that the fruits studied were longer (36.91 cm, approximately).

Regarding soluble solids (SS), significant differences were observed between maturity stages, increasing with it, observing 8.24 Brix in mature state, higher than the one reported (4.15 Brix) in Minas, Brazil [3]. In the composition of the ripe fruit (Table 2), it was observed that it has a low caloric intake with high water content (88.0 ± 0.1 g/100 g).

Table 1. Physicochemical characterization of the unripe, semi-ripe and ripe *Sicana odorifera* fruit pulp.

Variables	Unripe	Semi-Ripe	Ripe
Weight (g)	956 ± 21 [a]	2355 ± 12 [b]	1970 ± 51 [c]
Longitudinal diameter (cm)	19.76 ± 3.9 [a]	28.8 ± 0.7 [b]	26.9 ± 1.4 [b]
Transverse diameter (cm)	8.4 ± 1.1 [a]	11.2 ± 0.6 [b]	10.4 ± 0.7 [b]
Soluble solids (°Brix) *	3.26 ± 0.05 [a]	4.8 ± 0.1 [b]	8.2 ± 0.2 [c]
Titratable acidity (g of Ac. Citrus / 100 g) *	0.36 ± 0.08 [a]	0.22 ± 0.02 [b]	0.24 ± 0.04 [ab]
pH *	5.65 ± 0.02 [a]	6.37 ± 0.03 [b]	6.69 ± 0.04 [c]

The values are means ± SD. Different letters indicate significant differences between means (ANOVA and Tukey's a posteriori test, $p \leq 0.05$). * Determinations made in fresh fruit pulp (n = 3).

Table 2. *Sicana odorifera* ripe pulp composition.

Parameter	Value in Fresh Weight Pulp
Moisture (g/100 g)	88.0 ± 0.1
Ash (g/100 g)	0.15 ± 0.00
Total protein (g/100 g)	1.07 ± 0.08
Total carbohydrate (g/100 g)	5.55 ± 0.31
Dietary fiber (g/100 g)	2.92 ± 0.00
Caloric Value (Kcal/100 g)	26 ± 5

The values are means ± DS (n = 3). Determinations made in fresh fruit pulp.

3.2. Antioxidants Analysis

Statistically significant differences were observed in the contents of total phenols between the three maturity stages evaluated (Figure 1), unripe, semi-ripe and ripe (13.8 ± 0.8, 22.8 ± 2.2 and 37.5 ± 4.2 mg of GAE/100 g, respectively). The increase in total phenols as the fruit matures is important from the alimentary point of view, since it is the state in which it is consumed directly, and the health benefits associated with these compounds are expected. These results are also superior to those reported by Contreras-Calderón et al. [9] of fruits of *S. odorifera* harvested in Colombia (15.7 ± 1.1 mg of GAE/100 g) in fresh pulp and Marquez et al. [10] in fruits of another Cucurbitaceae, *Citrullus lanatus* (8.8 mg of GAE/100 g sample).

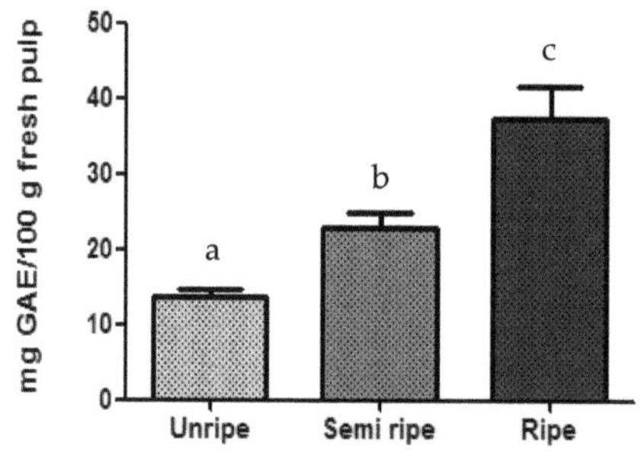

Figure 1. Total phenols content in unripe, semi-ripe and ripe *Sicana odorifera* fruit pulp. The bars represent the mean ± SD (n = 3). Different letters on the bars indicate significant differences between the means (ANOVA, Tukey's test a posteriori, $p \leq 0.05$).

The content of anthocyanins was higher (2.64 ± 0.10 mg/100 g) in the unripe fruit (Table 3), compared to the semi-ripe fruit (1.55 ± 0.70 mg/100 g) or ripe (not detectable); this behavior was contrary to what was observed in the content of total phenols, whose content increased with maturity (Figure 1). This could be due to the fact that the anthocyanins are stable at acid pH and the pH in the fruit of *Sicana odorifera* increased with maturity (Table 2). Anthocyanins are natural compounds present in a wide variety of plants, fruits and vegetables that are of great interest to the food industry, due to their qualities as colorants (given their range of colors from red to blue) and their antioxidant properties normally due to the presence of phenols in its structure. Although for *S. odorifera* no values of anthocyanins are reported, the content in other cucurbits such as watermelon is lower (0.4 mg/100 g) than observed in this work [10].

Regarding the vitamin C content, it was observed that it was higher in the ripe pulp (21.8 ± 4.7 mg/100 g) compared to the previous maturity stages. This result is higher than the value reported in pulp of *S. odorifera* harvested in Colombia (16.0 mg/100 g) reported by Contreras et al. [9] and Brazil (3.21 mg/100 g) reported by de Paula Filho et al. [3].

Table 3. Anthocyanins, vitamin C content and total antioxidant capacity (ABTS) in three stages of maturity in *Sicana odorifera* fruit pulp.

Parameter	Pulp		
	Unripe	Semi-Ripe	Ripe
Monomeric anthocyanins (mg/100 g of cyanide nidin 3-glucoside)	2.64 ± 0.10 [a]	1.55 ± 0.70 [a]	*
Vitamin C (mg/100g)	<14.30	<14.30	21.8 ± 4.7
Total antioxidant capacity ABTS (μM TEAC/g FW)	3.18 ± 0.39	6.75 ± 0.90	4.54 ± 0.24

The values are means ± SD. Different letters indicate significant differences between means (ANOVA and Tukey's test a posteriori, $p \leq 0.05$).

In the total antioxidant capacity determination by the ABTS method (Table 3), the behavior was different; the semi-ripe fruit (normally used as a vegetable in soups) presented a statistically higher value (6.75 ± 0.90 μM TEAC/g) in comparison with the values of unripe and ripe fruit (3.18 ± 0.39 and 4.54 ± 0.24 μM TEAC/g, respectively). The value observed in semi-ripe fruit is consistent with the value reported by Contreras et al. (2011) in ripe fruit (6.49 ± 0.47 μM TEAC/g) [9], although it is expected that there are differences between the chemical concentrations of fruits compared, due to the genetic variability in some criollo varieties of Cucurbitaceae. In the seeds and skin of the fruit, high contents of total phenols were observed (Table 4), and the total antioxidant capacity was higher (18–19 μmol of Trolox/100 g of ripe fruit peel).

Table 4. Total phenols content and total antioxidant capacity (ABTS) in *Sicana odorifera* fruits seeds and peel.

Variable	Seeds		Peel	
	Unripe	Ripe	Unripe	Ripe
Total phenols (mg GAE/100g FW)	66.9 ± 11.6 [a]	206 ± 27 [c]	80.7 ± 2.9 [a]	1689 ± 7 [d]
Total antioxidant capacity ABTS (μM TEAC/g FW)	4.44 ± 0.37 [a]	18.4 ± 1.6 [b]	12.2 ± 1.5 [c]	19.1 ± 0.3 [b]

The values are means ± SD. Different letters indicate significant differences between means (ANOVA and Tukey's a posteriori test, $p \leq 0.05$).

4. Conclusions

The *Sicana odorifera* fruits analyzed present an antioxidant potential of interest for the food industry, especially in its ripe and semi-ripe state, where the highest content of vitamin C and total phenols was observed, as well as the total antioxidant capacity (ABTS). This fruit can contribute to the supply of vitamin C in the diet in its fresh state and reduce the food insecurity of the population.

This work contributes to the scientific basis on the antioxidant potential of *S. odorifera* grown in Paraguay and opens a path towards the revaluation of the fruit. Continuing studies on its potential as a coloring and flavoring in the replacement of critical ingredients in foods, such as artificial additives, is a viable alternative based on its composition.

Acknowledgments: This work was supported by grant Ia ValSe-Food-CYTED (119RT0567). The authors are especially grateful to the "Kurugua poty" Foundation for the provision of the samples and the Facultad de Ciencias Químicas de la Universidad Nacional de Asunción for providing their facilities.

References

1. IICA. *Protocolos Estandarizados para la Valoración de Frutos Nativos del Procisur Frente a la Creciente Demanda por Ingredientes y Aditivos Especializados (Carotenoides, Antocianinas y Polifenoles)*; PROCISUR: Montevideo, Uruguay, 2018.
2. de Paula Filho, G.X.; Barreira, T.F.; Pinheiro, S.S.; Morais Cardoso, L.; Duarte Martino, H.S.; Pinheiro-Sant'Ana, H.M. 'Melão croá' (*Sicana sphaerica* Vell.) and 'maracujina' (*Sicana odorifera* Naud.): Chemical composition, carotenoids, vitamins and minerals in native fruits from the Brazilian Atlantic forest. *Fruits* **2015**, *70*, 341–349, doi:10.1051/fruits/2015035.
3. Lima, J.; Silva, M.P.; Teles, S.; Silva, F.; Martins, G. Avaliação de diferentes substratos na qualidade fisiológica de sementes de melão de caroá [*Sicana odorifera* (Vell.) Naudim]. *Rev. Bras. Plantas Med.* **2010**, *12*, 163–167, doi:10.1590/S1516-05722010000200007.
4. Zhan, Y.Y.; Wang, J.H.; Tian, X.; Feng, S.X.; Xue, L.; Tian, L.P. Protective effects of seed melon extract on CCl4-induced hepatic fibrosis in mice. *J. Ethnopharmacol.* **2016**, *193*, 531–537, doi:10.1016/j.jep.2016.10.006.
5. Mereles, L.; Ferro, E. Physical characrteristics, composition and minerals content in *Macadamia integrifolia* Maiden & Betche nuts, harvested in Cordillera Department, Paraguay. *Rojasiana* **2015**, *14*, 55–68.
6. Horwitz, W. (Ed.) *Official Methods of Analysis of AOAC International*, 17th ed.; AOAC International: Gaithersburg, MD, USA, 2000.
7. Singleton, V.L.; Rossi, J.A. Colorimetry of Total Phenolics with Phosphomolybdic-Phosphotungstic Acid Reagents. *Am. J. Enol. Vitic.* **1965**, *16*, 144–158. doi:10.12691/ijebb-2-1-5.
8. Re, R.; Pellegrini, N.; Proteggente, A.; Pannala, A.; Yang, M.; Rice-Evans, C. Antioxidant Activity Applying an Improved Abts Radical. *Free Radic. Biol. Med.* **1999**, *26*, 1231–1237, doi:10.1016/S0891-5849(98)00315-3.
9. Contreras, J.; Calderón, L.; Guerra, E.; García, B. Antioxidant capacity, phenolic content and vitamin C in pulp, peel and seed from 24 exotic fruits from Colombia. *Food Res. Int.* **2011**, *44*, 2047–2053, doi:10.1016/j.foodres.2010.11.003.
10. Marquez, L.; Torres, F.; Pretell, C. Antocianinas totales, fenoles totales y actividad antioxidante en pulpas de frutas Total anthocyanins and phenols and antioxidant activity in fruit pulps. *Pueblo Cont.* **2007**, *18*, 209–214.

proceedings

Proceedings

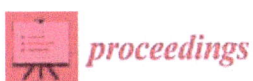

Microencapsulation of Sacha Inchi (*Plukenetia huayllabambana*) Oil by Spray Drying with Camu Camu (*Myrciaria dubia* (H.B.K.) Mc Vaugh) and Mango (*Mangifera indica*) Skins †

Rafael Alarcón [1], Billy Gonzales [1], Axel Sotelo [1], Gabriela Gallardo [2], María del Carmen Pérez-Camino [3] and Nancy Chasquibol [1,*]

1 Center of Studies and Innovation of Functional Foods (CEIAF)-Faculty of Industrial Engineering, Institute of Scientific Research, IDIC, Universidad de Lima, Avda. Javier Prado Este, 4600 Surco, Lima 33, Peru; ralarcor@ulima.edu.pe (R.A.); bgonzale@ulima.edu.pe (B.G.); alex_94sc@hotmail.com (A.S.)
2 Instituto Nacional de Tecnologia Industrial, INTI- Av. Gral Paz 5445, San Martín, Buenos Aires B1650, Argentina; ggallar@inti.gob.ar
3 Department of Characterization and Quality of Lipids, Instituto de la Grasa- CSIC, Ctra. Utrera km 1, Building 46, E-41013 Sevilla, Spain; mcperezcamino@ig.csic.es
* Correspondence: nchasquibol@ulima.edu.pe
† Presented at the 2nd Interational Conference of Ia ValSe-Food Network, Lisbon, Portugal, 21–22 October 2019.

Published: 26 August 2020

Abstract: Sacha inchi (*Plukenetia huayllabambana*) oil was microencapsulated by spray drying with gum arabic and with extracts of camu camu (*Myrciaria dubia* (HBK) Mc Vaugh) and mango (*Mangifera indica*) skins, obtained by assisted microwave. The physicochemical characteristics, such as moisture content, encapsulation efficiency, particle size, morphology, fatty acid composition and oxidative stability, were evaluated in order to select the best formulation for the development of functional foods. The most important results indicate that the microcapsules formulated with extracts of the fruit skins provide greater protection to sacha inchi oil (*P. huayllabambana*) against oxidation compared to commercial antioxidant BHT (Butylated Hydroxytoluene), resulting in a slight loss of ω-3 fatty acids.

Keywords: antioxidants; microencapsulation; assisted microwave extraction; oxidative stability; *Plukenetia huayllabambana*

1. Introduction

The camu camu (*Myrciaria dubia* (H.B.K.) Mc Vaugh) and mango (*Mangifera indica*) skins shown high antioxidant activity. Camu camu is a low-growing shrub found throughout the Amazon rainforest of Peru, Colombia, Venezuela and Brazil. The camu camu fruit is mainly consumed after being processed into juices, concentrates, and for the production of vitamin C capsules. As a result, a great volume of residue of seeds, skins and pulp that represent around 40% of the fruit in weight, are generated [1]. The importance of antioxidants is crucial for health, due to its ability to neutralize free radicals, which contain one or more unpaired electrons, being responsible for many degenerative diseases. Sacha inchi *Plukenetia huayllabambana* grows in the province of Rodríguez de Mendoza, Department of Amazonas-Peru, its oil contains a higher percentage of ω-3 (55.62 to 60.42% α-linolenic acid) [2]. However, an existing problem is that, due to its chemical structure, ω-3 acids have a high susceptibility to oxidation. A technology that emerges as an alternative to delay or inhibit its deterioration is microencapsulation [3], which consists in the preparation of an oil-in-water emulsion, containing encapsulating agents (or carriers, such as gums, fibers, proteins or carbohydrates) and their subsequent drying. This process aims to protect the poly-unsaturated fatty acids (PUFAs) from environmental factors, such as light, air or humidity. The aim of this work was to compare physicochemical characteristics and the oxidative stability of sacha inchi oil microcapsules elaborated with gum arabic and different mixtures of camu camu and mango skin extracts, in order to select the best formulation, which will allow for obtaining functional foods once the production of the microcapsules has escalated.

2. Materials and Methods

2.1. Raw Material

Cold pressed sacha inchi oil from the ecotype *P. huayllabambana* was obtained in the laboratory of *Centro de Estudios e Innovación del Alimento Funcional* (CEIAF) from the Universidad de Lima (Peru) and kept at 4 °C. Additionally, the camu camu and mango skins were washed and dried by infrared dryer (IRC DI8, Spain) at 40 °C, then ground into the food shredder (Grindomix GM200/Restch) and kept at −5 °C in polyethylene bags prior to phenolic extraction. The arabic gum (AG) was the agent encapsulant (wall material) due its versatility, good solubility, low viscosity at high concentrations and very good emulsifying properties.

2.2. Microwave-Assisted Extraction (MAE) of Polyphenols

The extraction of polyphenols from camu camu and mango dried skins were performed using a microwave oven (CW-2000, China) with ethanol:water. The optimal MAE conditions were previously determined. The resulting extracts were evaporated at 30 °C using a rotary evaporator (Büchi rotavapor R100 Labortechnic AG Switzerland).

2.3. Samples Proposed

A total amount of five samples to be analyzed was proposed, making mixtures of camu camu skin (CCS) extract, mango skin extract (MSE) with gum arabic (GA) and sacha inchi (*P. huayllabambana)* (SIPH) oil (Table 1).

Table 1. Samples of sacha inchi, *Plukenetia huayllabambana* oil microcapsules (SIPH) elaborated.

SIPH Oil Microencapsulated
SIPH + GA
SIPH + GA + CCSE[a] (220 ppm)
SIPH + GA + MSE[b] (220 ppm)
SIPH + GA + CCSE (110 ppm) + MSE (110 ppm)
SIPH + GA + BHT[c] (200 ppm)

[a] Camu camu skin extract, [b] Mango skin extract, [c] Commercial antioxidant.

2.4. Microencapsulation

The solutions were prepared with GA and distilled water containing 180 g of camu camu and mango skins extracts. Sacha inchi oil was then added at a concentration of 18% with respect to total solids. Emulsions were formed using a Silverson homogenizer L5M-A-England, operating at 9000 rpm for 10 m. The solutions were dried by spray dryer (Büchi B-290-Switzerland). Inlet and outlet air temperature were 140 °C and 70 °C, respectively, and the feed flow rate was 55 mL/min. The dried powders collected were stored in opaque hermetic bags at −5 °C for further analysis.

2.4.1. Moisture Determination

The moisture of the encapsulated samples was determined gravimetrically by drying until constant weight using halogen moisture analyzers (Sartorius MA-30, Germany), operating at 103 ± 2 °C.

2.4.2. Total Phenolic Content (TPC) and Surface Phenolic Content (SPC)

The TPC and SPC of the extract and powder was determined by the Folin–Ciocalteu method [4] with some modifications. The absorbance of the solution was measured at 760 nm using a spectrophotometer (1205 Vis Spectrophotometer UNICO). The results were expressed as μg of equivalent gallic acid (GAE) per gram of microcapsules (powder). All analyzes were done in triplicate. For the determination of the surface phenolic content (SPC), 24 mg of microcapsules were dissolved in 4.5 mL of methanol and stirred using a vortex for 1 min and then filtered through a Whatman filter paper number 2. The surface phenolic content was measured according to the same method described for TPC determination.

The percentage of efficiency of TPC microencapsulation was calculated using the following equation: Percentage Efficiency (%) = [(TPC) − (SPC)/TPC] × 100.

2.4.3. Determination of Antioxidant Activity on DPPH Radical

The determination of antioxidant activity was determined using DPPH as a free radical according the procedures described previously [5]. The absorptions of samples were then detected (Abs 517_{sample}). The percentage inhibition (% I) of free radicals was calculated using the following equation: Percentage Inhibition (% I) = [(Abs_{517} control) − (Abs_{517} sample)/(Abs_{517} control)] × 100.

2.4.4. Fatty Acid Composition

The fatty acid methyl esters (FAMEs) were prepared according to the International Union of Pure and Applied Chemistry, IUPAC [6] and the FAMEs formed were analyzed using a 7890B Agilent gas chromatograph (Agilent Technologies, Santa Clara, CA, USA) equipped with a SP2380 polar capillary column and a flame ionization detector (FID). The injector and detector temperatures were maintained at 225 and 250 °C, respectively. Hydrogen was used as carrier gas at a flow rate of 1.0 mL/min. The oven temperature was set at 165 °C and increased to 230 °C at 3 °C/min maintaining this temperature for 2 minutes. The injection volume was 1 μL.

2.4.5. Particle Size Distribution and Morphology

The particle size distribution was determined by laser diffraction spectroscopy on a Master Sizer Micro equipment (measuring range: 0.3 μm–300 μm). The average diameter of the equivalent volume or D [4,3] was informed. The photomicrographs were analyzed by a FEI scanning electron microscope, model QUANTA 250 FEG, (Hillsboro, OR, USA). Samples were previously gold sputtered with an Edwards Sputter Coater S150B (Crawley, England).

2.5. Oxidative Stability

The thermal analysis was carried out by differential scanning calorimetry (DSC), for the determination of oxidation onset temperature (OOT) according to the ASTM E2009-08 Standard Test Method for Oxidation Onset Temperature of Hydrocarbons by Differential Scanning Calorimetry.

3. Results and Discussion

According to Figure 1 the moisture content was between 3.74 and 4.32% being the highest value for the mixtures of antioxidants extracts, and the lowest values for the mango skin extract. The higher inlet (140 °C) and outlet temperature (80 °C) promoted the drying rate of droplets and resulted in low moisture content [7].

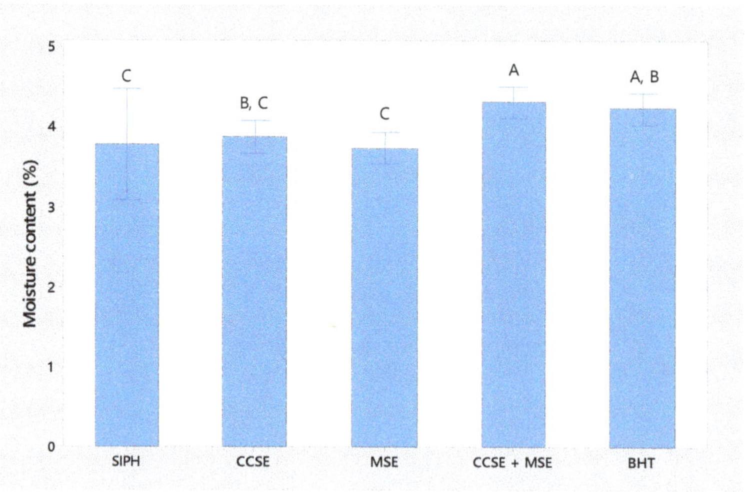

Figure 1. Moisture content (%) of sacha inchi, *P. huayllabambana* oil microencapsulated (SIPH). Different capital letters indicate significant differences ($p < 0.05$).

The total phenolic content (TPC) was found to be between 357.7 and 1677.9 μg GAE g^{-1} powder, and the highest amount of phenolic compound was obtained with camu camu and mango skin extract. There was a statistically significant difference in TPC between them. The encapsulation efficiency ranged from 90.25 to 98.28 %.

The microencapsulates showed the highest antioxidant activity (75.29 I% to 91.76 I%) and a composition very similar to the starting oil with a slight loss of omega-3 (57 vs. 58%) and a slight amount of *trans* fatty acid isomers (0.05–0.09%).

According to the particle size determination (Table 2), microcapsule diameters were between 1.6 and 20.9 μm. All samples showed a monodisperse distribution. The morphological analysis performed by SEM microscopy allowed us to distinguish rounded and concave microcapsules (Figure 2). This was an expected behavior for samples obtained by spray drying [8].

Table 2. D [4,3] values obtained from sacha inchi, *P. huayllabambana* oil microencapsulated (SIPH).

Formulation	D [4,3] μm	Span	Volume Distribution, μm		
			D(v, 0.1)	D(v, 0.5)	D(v, 0.9)
SIPH + GA	2.6 (0.1)	2.0 (6.1)	0.8 (0.1)	2.1 (0.1)	5.1 (0.1)
SIPH + GA + CCSE (220 ppm)	20.9 (1.4)	1.1 (0.1)	1.1 (0.1)	6.2 (0.1)	66.4 (0.2)
SIPH + GA + MSE (220 ppm)	1.6 (0.1)	1.4 (0.1)	0.7 (0.1)	1.4 (0.1)	2.7 (0.2)
SIPH + GA + CCSE (110 ppm) + MSE (110 ppm)	4.0 (1.4)	1.5 (0.1)	0.7 (0.1)	1.4 (0.1)	2.9 (0.2)

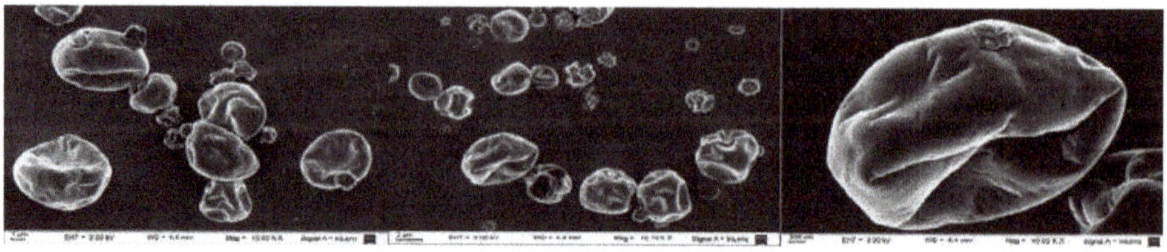

Figure 2. Morphology of microcapsule: SIPH + GA + CCSE (110 ppm) + MSE (110 ppm).

The results obtained for the oxidative stability of microcapsule with naturals antioxidants were to be between 188 to 198 °C and for the commercial antioxidant were 174 °C, thus showed that the antioxidant extracts from natural origin provided greater protection to sacha inchi oil (*P. huayllabambana*) against oxidation.

4. Conclusions

The microcapsules of sacha inchi (*P. huayllabambana*) oil showed a low percentage of moisture content (3.74 to 4.32%), a high amount of phenolic compound (357.7 to 1677.9 μg GAE g^{-1} power) and a high encapsulation efficiency (90.25 to 98.28 %). The microencapsulates showed high antioxidant activity (75.29 I% to 91.76 I%) with a composition very similar to the starting oil with a slight loss of omega-3 (57 vs. 58%) and a slight amount of *trans* fatty acid isomers (0.05 to 0.09%). The morphological analysis allowed us to distinguish rounded and concave microcapsules, and they vary in their shape due to their chemical composition. The results obtained from OOT showed that the antioxidant extracts from natural origin provide greater protection to sacha inchi oil (*P. huayllabambana*) against oxidation. For this reason, the microcapsules can be used as a natural antioxidant in functional foods or nutraceutical products, with possible health benefits.

Acknowledgments: This research is part of the project N° 093-INNOVATEPERU-IDIBIO "Development of functional drink, source of omega-3 and antioxidants microencapsulated from camu camu and mango skins, to promote the commercial development of Peruvian biodiversity" the INNOVATE PERU, belonging to the Ministry of Production—Peru, and Universidad de Lima—Peru. The authors want to thank Ia ValSe-Food-CYTED (119RT0567) for financing the participation in the CYTED meeting.

References

1. Rodríguez, R.; Menezes, H.; Cabral, I.; Dornier, M.; Reynes, M. An Amazonian fruit with a high potential as a natural source of vitamin C: The camu camu (*Myrciaria dubia*). *Fruits* **2001**, *56*, 345–354. Available online: https://hal.archives-ouvertes.fr/hal-01517199/document (August 18, 2019)
2. Chasquibol, N.; Moreda, W.; Yácono, J.C.; Guinda, A.; Gómez-Coca, R.; Pérez-Camino, M.C. Characterization of Glyceridic and Unsaponifiable Compounds of Sacha Inchi (*Plukenetia huayllabambana* L.) Oils. *J. Agric. Food Chem.* **2014**, *62*, 10162–10169, doi:10.1021/jf5028697
3. Carneiro, H.; Tonon, R.; Grosso, C.; Hubinger, M. Encapsulation efficiency and oxidative stability of flaxseed oil microencapsulated by spray drying using different combinations of wall materials. *J. Food Eng.* **2013**, *115*, 443–451, doi:10.1016/j.jfoodeng.2012.03.033
4. Saénz, C.; Tapia, S.; Chávez, J.; Robert, P. Microencapsulation by spray drying of bioactive compounds from cactus pear (*Opuntia ficus-indica*). *Food Chem.* **2009**, *114*, 616–622, doi:10.1016/j.foodchem.2008.09.095
5. Luo, W.; Zhao, M.; Yang, B.; Shen, G.; Rao, G. Identification of bioactivecompounds in *Phyllenthus emblica* L. fruit and their free radical scavengingactivities. *Food Chem.* **2009**, *114*, 499–504, doi:10.1016/j.foodchem.2008.09.077
6. IUPAC. Standard Method 2.301. *Standard Methods for the Analysis of Oils, Fats and Derivatives. Preparation of Fatty acid Methyl Ester*; Blackwell Scientific: Oxford, UK, **1987**.
7. Rajam, R.; Karthik, P.; Parthasarathi, S.; Joseph, G.S. Anandharamakrishnan C. Effect of whey protein e alginate wall systems on survival of microencapsulated Lactobacillus plantarum in simulated gastrointestinal conditions. *J. Func. Food* **2012**, *4*, 891–898, doi:10.1016/j.jff.2012.06.006
8. Finney, J; Buffo, R; Reineccius, G. A. Effect of type of atomization and processing temperatures on the physical properties and stability of spray-dried flavors. Journal of Food Science. **2002**, *67*(3):1108-14, doi:10.1111/j.1365-2621.2002.tb09461.x

proceedings

MDPI

Proceedings

Changes in the Antioxidant Activity of Peptides Released during the Hydrolysis of Quinoa (*Chenopodium quinoa* willd) Protein Concentrate †

Julio Rueda, Manuel Oscar Lobo and Nornma Cristina Sammán *

Centro de Investigaciones Interdisciplinarias en Tecnologías y Desarrollo Social para el NOA (CIITED), CONICET-Facultad de Ingeniería, Universidad Nacional de Jujuy, Ítalo Palanca 10.4600, San Salvador de Jujuy, Argentina; julioruedafca@gmail.com (J.R.); mlobo958@gmail.com(M.O.L.)

* Correspondence: normasamman@gmail.com

† Presented at the 2nd International Conference of Ia ValSe-Food Network, Lisbon, Portugal, 21–22 October 2019.

Published: 26 August 2020

Abstract: There is an increased interest in Andean crops as sources of nutritious compounds. This study evaluated changes in the antioxidant activity of quinoa protein hydrolysate with commercial enzymes. Aliquots at 0, 30, 60, 120 and 180 min were tested for DPPH (2,2′-diphenyl-1-picrylhydrazyl) and ABTS ((2,2′-azino-bis(3-ethylbenzothiazoline-6-sulfonic acid)), antiradical activity. Initial DPPH inhibition rose from 9.2% ± 2.7% to 20.0% ± 4.0% (30 min) when employing alcalase and initial ABTS inhibition increased from 20.9 ± 0.2 to 105.1 ± 3.7 with ascorbic acid μg/mL (30 min). Protamex improved this to 75.7 ± 0.6 μg/mL (180 min). Alcalase and protamex are suitable enzymes for the production of rich peptides and hydrolysates as novel ingredients with antioxidant activity.

Keywords: enzymes; hydrolysate; peptides; protein; quinoa

1. Introduction

Several plant-derived foods exhibit health benefiting attributes and are suitable for healthy food production. Many of these properties are attributed to proteins and peptides. Peptides can be found in foods as individual parts of proteins or encrypted inside parent proteins.

Peptides and hydrolysates are produced from diverse protein sources. Although animal sources such as milk, eggs and meat proteins are the largest type of products employed, they are not cheap or easily accessible. This has led to an increased interest in vegetable proteins for the manufacturing of such products.

Quinoa has been largely consumed by early Latin American inhabitants and has a long tradition of well-known nutritive properties, now appreciated by different regions around the world [1]. There are few studies showing the potentiality of quinoa protein as substrate for the release of bioactive peptides or hydrolysate as novel ingredients [2]. The scope of this study was to evaluate the potentiality of quinoa for the production of protein concentrate and to select widely available enzymes for the production of protein hydrolysates as functional ingredients with antioxidant activity.

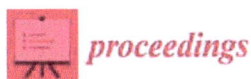

2. Materials and Methods

2.1. *Chemicals and Reagents*

Chemical and reagents employed were of analytical grade. The following were purchased from Sigma-Aldrich: 2,4,6-trinitrobenzenesulfonic (TNBS) acid solution (5%), L-leucine (≥98% HPLC), alcalase from *Bacillus lincheniformis* (activity ≥ 2.4 AU/g), flavourzyme from *Aspergillus oryzae* (activity ≥ 500 LAPU/g) and protamex from *Bacillus* sp. (activity ≥ 1.5 AU-NH/g). AU is defined as Anson units; LAPU is defined as leucine aminopeptidase units[3]. Reagents 2,2'-diphenyl-1-picrylhydrazyl (DPPH) and 2,2-azino-bis-3-ethylbenzothiazoline-6-sulfonic acid (ABTS) were obtained from Merck Bio and Sigma-Aldrich, respectively. Quinoa seeds (*Chenopodium quinoa* willd. var INTA Hornillos) were provided by the National Institute of Agricultural and Livestock Technology (INTA-IPAF-NOA-Argentina). The pH was adjusted using NaOH or HCl 0.3 M. Assays were performed in duplicate (n = 2). A two-way ANOVA and the Tukey test (Graph Pad Prism 6.0.) were performed. Significant differences were detected at $p < 0.05$.

2.2. *Quinoa Protein Solubility Study*

2.2.1. Quinoa Protein Solubility

Quinoa grains were ground in a mill and the flour was agitated for 2 h with petroleum ether (1:10 *w*/*v*). Protein solubility at different pH levels (212) was determined. Aqueous quinoa flour suspensions (1/10 *w*/*v*) were mixed for 20 min, at 30°C and 100 rpm. Proteins were measured at pH 2–12 using bovine serum albumin (BSA) as standard and Bradford's reagent. Results are expressed as mg BSA equivalents/mL.

2.2.2. SDS-PAGE

Electrophoresis in denaturing conditions was performed using acrylamide stacking (4%) and running (12%) gels. Samples were boiled for 3 min in separating buffer containing 2% SDS, 10% glycerol, 0.01% bromophenol blue, 0.0625 M tris-HCl pH 6.8 and 5% β-mercaptoethanol, loaded (5 μL) onto gels and run at a constant voltage (60 V first 10 min and 120 V) using Laemmli buffer. Gel staining was performed with Coomassie Blue R-250 and destaining was performed in methanol/acetic acid solution (50/20).

2.3. *Quinoa Protein Concentrate (QPC)*

A proportion of 1:10 (*w*/*v*) of defatted flour was agitated at 150 rpm, for 2 h, at 30 °C, at the desired pH, then centrifuged (10,000× *g*, 10 min), and proteins were precipitated at acid pH. The slurry was refrigerated for 30 min, at 4 °C, and centrifuged (10,000× *g*, 10 min, 4 °C). The pellet was air-dried in a flux oven (30 °C, 12 h). The protein concentrate was powdered and the protein content (N × 5.7) was determined by the Kjeldahl method.

2.4. *Proteolysis*

2.4.1. Hydrolysis Conditions

Hydrolysis conditions of quinoa protein (10 mg/mL) at pH 7–10 and temperatures of 40–60 °C were determined. Proteases were added (1/10 w or *v*/*w*) and after 10 min, the reaction was stopped (1.0 mL TCA 10%). A blank for each assay was prepared without the enzyme. The slurry was refrigerated (4 °C), for 30 min and centrifuged (15,000× *g*, 4 °C and 10 min). TCA soluble peptides were determined.

2.4.2. Protease Activity

Alpha amino groups were measured according [4]. Aliquots of 0.1 mL of peptides were mixed with 3.4 mL phosphate buffer pH 8.2 (0.2 M) and 0.5 mL of TNBS 0.05% and incubated in the dark at

50 °C, for 60 min at 200 rpm. The absorbance was measured at 420 nm in a UV-visible spectrophotometer, using leucine as standard. Proteolytic units were expressed as leucine equivalents mM/min (PU/min) [3].

2.4.3. Quinoa Protein Hydrolysates (QPH)

Enzyme and substrate ratio was 1:10. Optimal conditions were adjusted. Aliquots were taken at 0, 30, 60, 120 and 180 min. Enzyme inactivation was performed at 85 °C for 10 min.

2.4.4. Hydrolysis Degree Calculation

The hydrolysis degree percentage (HD%) was determined as follows:

$$HD\% = B \times N_b \times \frac{1}{\alpha} \times \frac{1}{PM} \times \frac{1}{h_{tot}} \times 100$$

h_{tot} is the total number of peptide bonds (7.21 miliequivalents/g protein), calculated on the basis of the amino acid occurrence in a local variety of quinoa [5], B is the volume (mL) of base necessary to keep the pH constant, N_b is the normality of the base, α^{-1} is the calibration factor calculated as the reciprocal of the average degree of dissociation of α-NH amino groups, and PM is the mass of protein (g) in the total reaction.

2.5. Antiradical Activity

2.5.1. Inhibition of Radical DPPH

QPH antiradical properties were tested following the methodology of Chakka et al. [6]. Briefly, 0.1 mL of sample or blank (distillate water) was mixed with 1.4 mL of DPPH at 0.1 mM in anhydrous methanol and incubated 30 min in the dark; absorbance readings were taken at 515 nm in a UV-visible spectrophotometer. The antiradical activity as percentage (ARA) was calculated according the following equation.

$$ARA\ (\%) = \left[\frac{1 - (A_s - A_b)}{A_c}\right] \times 100$$

A_c, A_s and A_b are absorbance of control, sample and blank, respectively.

2.5.2. Inhibition of Radical ABTS

The radical formed with 2,2′-azinobis (3-ethylbenzothiazoline-6-sulfonic acid) ammonium salt and potassium persulfate was employed. The radical solution was mixed with 50 μL of sample or standard and incubated for 6 min before absorbance reading. A linear calibration curve was prepared using ascorbic acid. Results are expressed as equivalent ascorbic acid μg/mL.

3. Results and Discussion

3.1. Quinoa Protein Solubility

Figure 1 shows the protein solubility in a wide range of pH levels; an increase in solubility profile at extreme acid and alkali conditions was observed. The maximal concentrations were measured at pH 9 (13.33 ± 0.71) and 10 (13.41 ± 0.62). Similar results were reported by Elsohaimy et al. [7]. In other Andean crops such as amaranth, pH values of 4 or 5 precipitated most proteins [8], which agrees with this study. The protein content of the concentrate was 61.61% ± 0.43%. This value is higher than the one (40.7% ± 0.9%) found by Nongonierma et al. [2], who employed mashed grains, an extraction time of 60 min compared to 120 min and quinoa defatted flour as starting material, as described in this study. The higher value may be due to the much smaller and homogeneous material employed, which allowed more protein extractability. Higher levels of protein content (65.5 ± 0.1 and 77.2 ± 0.1% *w/w*) were found in protein concentrates prepared by Aluko and Monu [9] and Abugoch et al. [10].

3.2. Electrophoresis

Figure 2 shows the electrophoretic run of proteins solubilized at pH 9 (lane 2) and 4 (lane3). While lane 3 shows no protein bands, indicating that the solubilized protein meets its isoelectric point, the pattern of proteins solubilized at pH 9 (lane 2) exhibits polypeptides of high molecular weight, ranging from 100.0 to 24.5 kDa. Major bands of polypeptides are estimated in molecular weights of 45,700 (A), 32,000–28,700 (B) and 25,900–24,480 (C) kDa. These sets of polypeptides have been initially described by Brinegar and Goudan [11] and recently by Vilcacundo et al. [1].

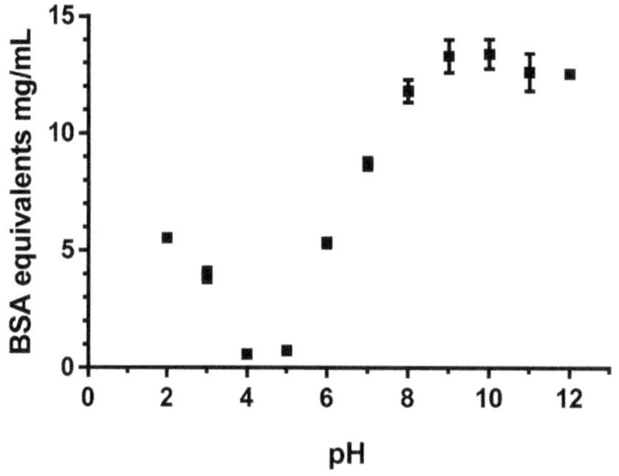

Figure 1. Quinoa protein solubility at different pH values. Obtained from aqueous suspensions of defatted flour (1:10 *w/v*) maintained at a constant pH (NaOH or HCl) for 10 min at 30 °C and 100 rpm. n = 2. Values (mean ± SD) are expressed as bovine serum album equivalents mg/mL.

The polypeptides belong to the main constituent protein in quinoa, chenopodin. Each group (B and C) is composed of subunits bonded by disulfide bonds. In Figure 2, bands B and C correspond respectively to the acidic and basic polypeptides found in 11S-type globulin family proteins. The 45,000 kDa protein appearing in Figure 2 may be a chenopodin A-B protein with a strong disulfide bond still remaining after reductive SDS-PAGE conditions; bands appearing under 20 kDa have been described as 2S type proteins present in many seeds, albumins mainly. Concerning amino acid composition, it has been reported that chenopodin exceeds the Food and Agriculture Organization(FAO) requirements and has a high chemical score [11].

3.3. Hydrolysis Conditions

Figure 3 shows the leucine equivalents at different conditions of pH and temperature, using quinoa protein as substrate. Figure 3B (flavourzyme) and Figure 3C (protamex) show an increase in the protease activity towards neutral conditions for all temperatures and pH levels. Flavourzyme showed no activity at pH 9–10, at 60 °C. On the other hand, alcalasa (Figure 3A) was very active at all pH levels and temperatures. Its activity increased noticeably as the medium turned alkaline. This increase was gradual at 40 °C but high and constant at 50 or 60 °C. Optimal conditions and proteolytic units (PU) are summarized in Table 1, solubilized from defatted flour 1:10 *w/v* in aqueous suspensions. Alkali and acid media were regulated with NaOH or HCl.

The optimal values of temperature found (flavourzyme and protamex) differ from those employed by Jung et al. [12]. This change could be attributed to the substrate type, enzyme: substrate ratio and conditions of reaction employed for the hydrolysis.

Table 1. Experimental proteolytic units (PU) and optimal temperature and pH for the hydrolysis of quinoa protein concentrate.

Enzyme	Temperature	pH	PU/min
Alcalasa	50	8	220
Flavoruzyme	60	7	125
Protamex	50	7	182

PU/min: Proteolytic units expressed as leucine equivalents mM/min.

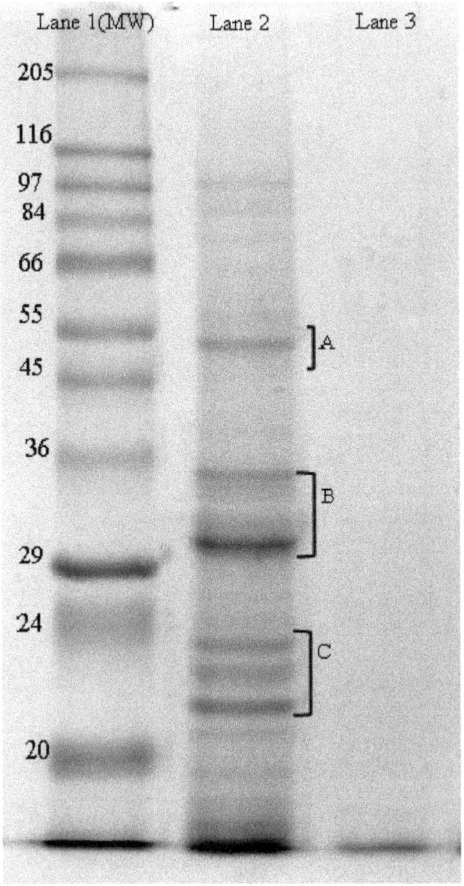

Figure 2. SDS-PAGE characterization of quinoa proteins soluble at pH 9 (lane 2) and pH 4 (lane 3). MW: molecular weight marker; A: 7S globulin 45 kDa; B: globulin acid subunit; C: globulin basic subunit.

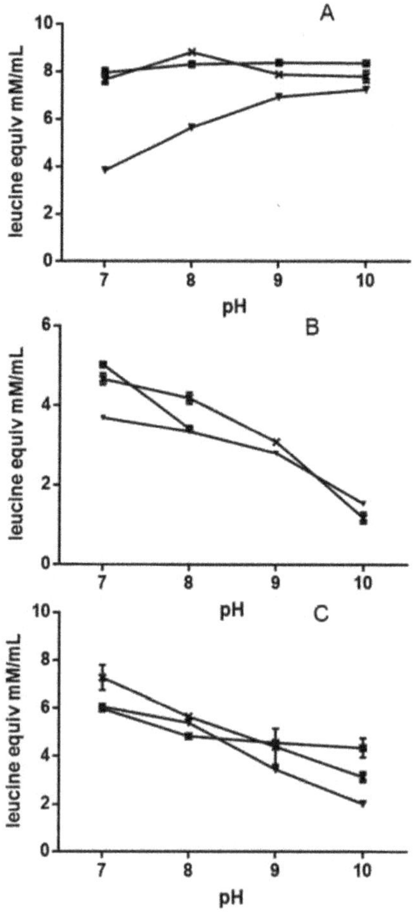

Figure 3. Leucine equivalents produced at different pH levels and temperatures by three enzymes (substrate enzyme ratio 1:10). (**A**) alcalase; (**B**) flavourzyme; (**C**) protamex. Temperatures: 40 °C (▼); 50 °C (×); 60 °C (■). Results (mean ± SD) are expressed as mM/mL of supernatant after 10 min of hydrolysis (n = 2).

3.4. Quinoa Protein Hydrolysis

As shown in Figure 4, the hydrolysis of quinoa protein follows an enzyme dependent hydrolysis pattern. Curves show a high rate of hydrolysis in the first 30 min with alcalsa (27.9%) and protamex (20.7%) and evolves at low velocity for flavourzyme (4.3%). Thamnarathip et al. [13] achieved 13–14 HD% by employing alcalasa, flavourzyme and rice bran protein after 6 h.

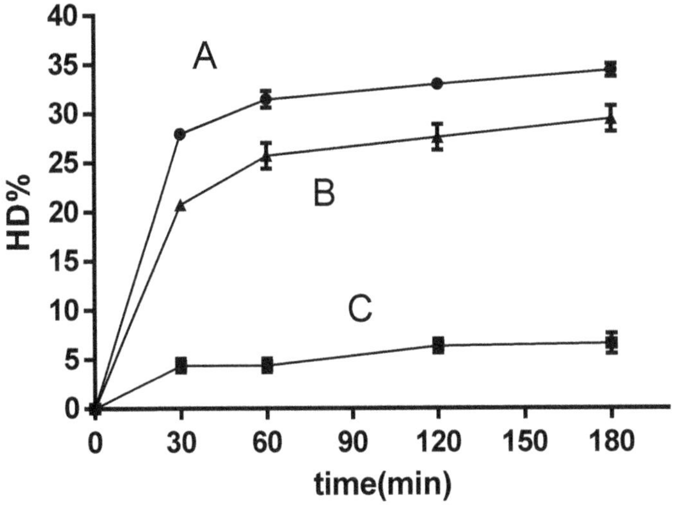

Figure 4. Hydrolysis degree progression of quinoa protein concentrate in aqueous suspension (10 mg/mL) using different enzymes (substrate enzyme ratio 1:10). Data are expressed as percentage (mean ± SD) of peptide bonds cleaved at defined time (min).

3.5. Antiradical Activity

Table 2 compares the antiradical activity measured by ABTS and DPPH methods as the hydrolysis progresses. DPPH antiradical activity with alcalasa was higher than that obtained with flavourzyme and protamex. From initial values of 9.3% ± 0.2%, alcalase increased activity to 20.0% ± 4.0% after 30-min hydrolysis. DPPH inhibition with flavourzyme and protamex decreased or remained fairly constant as the hydrolysis degree increased, with 14% being the average highest value after 180 min.

ABTS measurements were remarkably different among the enzymes employed. From initial values of 20.9 ± 0.2 ascorbic acid μg equivalents/mL of the non-hydrolyzed protein, the antioxidant activity increased to 105.1 ± 0.1 (30 min) and to 75.7 ± 0.6 μg/mL (180 min) using alcalasa and protamex, respectively. Flavourzyme reduced the antioxidant activity. Differences in the activities measured by the mentioned methods may be related to the solubility of the antioxidant compounds in the organic (methanol) and aqueous phases employed. This could be related to lipophilic and hydrophilic peptides and also to the low molecular weight.

Table 2. Antiradical activity of quinoa protein hydrolyzates at different times of hydrolysis.

	Alcalase				Flavourzyme				Protamex			
Time (min)	**ABTS (asc. ac./mL)**		**DPPH (% inhib)**		**ABTS (asc. ac./mL)**		**DPPH (% inhib)**		**ABTS (asc. ac./mL)**		**DPPH (% inhib)**	
0	20.9	±0.2c	9.3	±2.7a	20.9	±0.2a	9.3	±2.7a	20.9	±0.2d	9.3	±2.7a
30	105.1	±3.7b	20.0	±4.0b	10.9	±0.3b	9.6	±1.0a	42.8	±1.4c	11.3	±3.0a
60	110.9	±1.3ab	11.78	±1.4ab	15.1	±0.5b	9.4	±0.5a	50.0	±1.2c	12.3	±3.3a
120	102.9	±2.4b	10.8	±2.5a	15.3	±0.6b	4.0	±0.9a	60.1	±2.6b	14.1	±0.7a
180	119.19	±8.7a	7.6	±6.2a	12.7	±2.0b	14.8	±2.4b	75.7	±0.6a	12.5	±1.3a

ABTS expressed as ascorbic acid μg equivalents/mL of hydrolyzate; DPPH expressed as percentage of inhibition of 0.1 mL of hydrolyzate. Values followed by the same letter within each column are not significantly different, $p > 0.05$.

4. Conclusions

Results reported here show quinoa as a promising source for the production of protein concentrates and protein hydrolysates with potential antioxidant activity and the formulation of novel food ingredients with bioactive properties. Moreover, commercial enzymes were assayed from the hydrolysis of quinoa protein and the release of peptides, alcalasa and protamex being preferable for the hydrolysis of chenopodin at high rates in practically short times of reaction. Additional work needs to be performed in order to characterize the peptides responsible for the potential bioactivity.

Acknowledgments: This work was supported by grant Ia ValSe-Food-CYTED (119RT0567), Consejo Nacional de Investigaciones Científicas y Técnicas (CONICET) and Secretaría de Ciencia y Técnica y Estudios Regionales (SECTER), Universidad Nacional de Jujuy, Argentina.

References

1. Vilcacundo, R.; Martínez-Villaluenga, C.; Hernández-Ledesma, B. Release of dipeptidyl peptidase IV, α-amylase and α-glucosidase inhibitory peptides from quinoa (*Chenopodium quinoa* Willd.) during in vitro simulated gastrointestinal digestion. *J. Funct. Foods* **2017**, *35*, 531–539, doi:10.1016/j.jff.2017.06.024.
2. Nongonierma, A.B.; Le Maux, S.; Dubrulle, C.; Barre, C.; FitzGerald, R.J. Quinoa (*Chenopodium quinoa* Willd.) protein hydrolysates with in vitro dipeptidyl peptidase IV (DPP-IV) inhibitory and antioxidant properties. *J. Cereal Sci.* **2015**, *65*, 112–118, doi:10.1016/j.jcs.2015.07.004.
3. Navarrete del Toro, M.A.; García-Carreño, F.L. Evaluation of the Progress of Protein Hydrolysis. *Curr. Protoc. Food Anal. Chem.* **2003**, 10, B2.2.1–B2.2.14, doi:10.1002/0471142913.fab0202s10.
4. Schokker, E.P.; van Boekel, A.J.S. Kinetic Modeling of Enzyme Inactivation: Kinetics of Heat Inactivation at 90–110 °C of Extracellular Proteinase from Pseudomonas fluorescens 22F. *J. Agric. Food Chem.* **1997**, *45*, 4740–4747, doi:10.1021/jf970429i.
5. Mota, C.; Santos, M.; Mauro, R.; Samman, N.; Matos, A.S.; Torres, D.; Castanheira, I. Protein content and amino acids profile of pseudocereals. *Food Chem.* **2016**, *193*, 55–61, doi:10.1016/j.foodchem.2014.11.043.
6. Chakka, A.K.; Elias, M.; Jini, R.; Sakhare, P.Z.; Bhaskar, N. *In-vitro* antioxidant and antibacterial properties of fermentatively and enzymatically prepared chicken liver protein hydrolysates. *J. Food Sci. Technol.* **2015**, *52*, 8059–8067, doi:10.1007/s13197-015-1920-2.
7. Elsohaimy, S.A.; Refaay, T.M.; Zaytoun, M.A.M. Physicochemical and functional properties of quinoa protein isolate. *Ann. Agric. Sci.* **2015**, *60*, 297–305, doi:10.1016/j.aoas.2015.10.007.
8. Martínez, E.N.; Añón, M.C. Composition and structural characterization of amaranth protein isolates. An electrophoretic and calorimetric study. *J. Agric. Food Chem.* **1996**, *44*, 2523–2530, doi:10.1021/jf960169p.
9. Aluko, R.E.; Monu, E. Functional and Bioactive Properties of quinoa seed protein hydrolysates. *Food Chem. Toxicol.* **2003**, *68*, 1254–1258, doi:10.1007/s10577-010-9145-8.
10. Abugoch, L.E.; Romero, N.; Tapia, C.A.; Silva, J.; Rivera, M. Study of some physicochemical and functional properties of quinoa (*Chenopodium quinoa* Willd) protein isolates. *J. Agric. Food Chem.* **2008**, *56*, 4745–4750, doi:10.1021/jf703689u.
11. Brinegar, C.; Goundan, S. Isolation and characterization of chenopodin, the 11S seed storage protein of quinoa (*Chenopodium quinoa*). *J. Agric. Food Chem.* **1993**, *41*, 182–185, doi:10.1021/jf00026a006
12. Jung, K.; Karawita, R.; Heo, S.J.; Lee, B.J.; Kim, S.K.; Jeon, Y.J. Recovery of a novel Ca-binding peptide from Alaska Pollack (*Theragra chalcogramma*) backbone by pepsinolytic hydrolysis. *Process Biochem.* **2006**, *41*, 2097–2100, doi:10.1016/j.procbio.2006.05.008.
13. Thamnarathip, P.; Jangchud, K.; Jangchud, A.; Nitisinprasert, S.; Tadakittisarn, S.; Vardhanabhuti, B. Extraction and characterisation of Riceberry bran protein hydrolysate using enzymatic hydrolysis. *Int. J. Food Sci. Technol.* **2016**, *51*, 194–202, doi:10.1111/ijfs.13008.

proceedings

MDPI

Proceedings

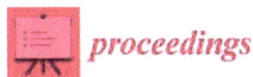

Technological and Sensory Properties of Baby Purees Formulated with Andean Grains and Dried with Different Methods †

María Dolores Jiménez, Manuel Oscar Lobo and Norma Cristina Sammán *

Faculty of Engineering- CIITED CONICET, National University of Jujuy, San Salvador de Jujuy 4600, Argentina; doloresjimenez4@gmail.com (M.D.J.); mlobo958@gmail.com@gmail.com (M.O.L.)

* Correspondence: normasamman@gmail.com

† Presented at the 2nd International Conference of IaValSe-Food Network, Lisbon, Portugal, 21–22 October 2019.

Published: 26 August 2020

Abstract: The aim of this work was to compare different cooking–drying methods to obtain dehydrated baby purees. Flours of quinoa and amaranth (native and germinated) were used to formulate them. Dry powders (DPs) were obtained by lyophilization (LD), convection (CD), and extrusion (ED). Proximal composition, particle size and morphology, water absorption capacity, and solubility were evaluated in DPs. Color, texture profile (TP), and sensory characteristics were determined in fresh pure and rehydrated powders (RPs). The LD particles were smaller and homogeneous; CD showed collapsed particles, and ED presented agglomerated particles. Different drying methods influenced the rehydration properties of DPs, as well as the color, TP, and sensory evaluation of RPs. The best method to obtain dehydrated baby purees was extrusion.

Keywords: amaranth; Andean potato; dehydrated powder; germination; puree; quinoa

1. Introduction

Complementary feeding begins after the first six months of life for babies. The feeding in this stage is fundamental for their physical and mental development. The incorporated foods must have semi-solid consistency and be very digestible [1].

The quinoa and amaranth Andean grains are free of gluten, have proteins of high biological value, and are rich in minerals, vitamins, fiber, and antioxidant compounds. During germination of grains, proteins, lipids, and starches are hydrolyzed, and the content of antioxidant compounds improves. The germinated grains are suitable for the formulation of baby foods due to their greater digestibility with respect to the native grains; however, the modifications that occur during germination influence the thermal, rheological, textural, and sensory characteristics of the final product [2].

The equilibrium humidity of dehydrated foods ensures their microbiological, chemical, and enzymatic stability over a prolonged period. It is important to determine the appropriate drying method for each type of food, because the variables of the process influence the nutritional, technological, and sensory characteristics of the dehydrated and rehydrated food [3,4].

The aim of this work was to formulate a dehydrated puree to reconstitute for babies that is made with quinoa and amaranth flours (germinated and non-germinated).

2. Materials and Methods

2.1. Raw Material

Quinoa (*Cica*), amaranth (*Mantegazzianus*), and Andean potato (*Collareja*) were obtained from Centro de Investigación y Desarrollo Tecnológico para la Agricultura Familiar, Hornillos, Jujuy-Argentina.

2.1.1. Mashed Potato

Potatoes were washed, cooked in boiling water, peeled, and mashed.

2.1.2. Non-Germinated and Germinated Grain Flours

The grains were washed, and the saponin of quinoa was removed. A part of the washed grains was soaked in tap water (6 h, at room temperature) and sprouted (22–24 °C, 80–90% RH, in darkness) 24 h quinoa and 48 h amaranth. Then, they were dried in a forced circulation oven (50 °C) and milled.

2.2. Fresh Puree and Dry Powders

2.2.1. Fresh Puree (FP)

Mixture of mashed Andean potato (8.0 g), flours of quinoa (5.0 g), germinated quinoa (2.0 g), amaranth (3.0 g), and germinated amaranth (1.0 g) with dry milk (3.0 g), sugar (3.0 g), and tap water (81 mL) were cooked (30 min). Banana essence (400 μL) and citric and ascorbic acids (0.025 and 0.015 mL, respectively) were added. The cooked puree was packed in glass flasks with screwed metal caps and autoclaved (119 °C, 15 min).

2.2.2. Freeze-Drying (LD)

The cooked puree was frozen at −70 °C, and then freeze-dried at −70 °C and pressure of 0.15 mbar during 24 h, in Heto FD4 (Heto-Holten, Denmark).

2.2.3. Drying by Forced Air Circulation (CD)

It was carried out by using a convective dryer Memmert Radiant Warmer Model A52200-35_Vac 230 (Memmert, Schwabach, Germany). The cooked puree was distributed with a thickness of 1.9 mm on a tray and dried (40 °C, 12 h). An air flow of 55 ft^3/h was used.

2.2.4. Extrusion (ED)

The mixture of the flours with the other ingredients were extruded, using a twin-screw extruder DT65 (Incalfer, Buenos Aires, Argentina), with a feed rate of 12 kg/h, water flow of 100 cc/min, and screw speed of 1500 rpm (die diameter 3.0 mm). Extrusion was carried out with three temperature sectors (45, 175, and 180 °C).

2.2.5. Dry Powders (DPs)

They were obtained from fresh puree by freeze-drying, forced air circulation, and extrusion. The samples were milled in a centrifugal mill CHINCAN model FW 100 (CHINCAN, Hangzhou, China), vacuum-packed in polyethylene bags, and stored at room temperature.

2.3. Proximal Composition

It was determined by official techniques AOAC [5] moisture (method 925.10), ash (method 923.03), lipid (Soxhlet method), and total nitrogen (method 920.87). Nitrogen–protein conversion factors 6.25 was used. Carbohydrate was determined by difference.

2.4. Particle Morphology

Particle morphology was studied by scanning electron microscope (SEM), using SEM Zeiss Supra model 55 VP (Zeiss, Oberkochen, Germany). The powders were placed on a specimen holder, with the help of double-sided Scotch Tape and sputter-coated with gold (2 min, 2 mbar), and finally each sample was transferred to a microscope, where it was observed (15 kV, 9.75 × 10^{-5} Torr).

2.5. Rehydration Properties

2.5.1. Solubility and Water Absorption Capacity

The solubility and water absorption capacity (WAC) were studied according to Wani et al. [6]. DP was weight in a tared centrifuge tube, and water was added. It was mixed and then centrifuged. The supernatant was recovered and dried in a convection oven. The dry solid was weight, and the solubility calculated as g soluble solids/100 g product. The precipitate was weighed, and WAC was calculated as g water absorbed/g product.

2.5.2. Water Adsorption

It was determined according to Tonon et al. [7]. DP was weight (2.2 g) in capsules and placed in a chamber (25 °C, 76% RH achieved with 6 molal NaCl solution). Water activity and moisture were registered. The experimental data were modeled with the BET equation (Equation (1)):

$$\text{BET (Brunauer, Emmett, and Teller):} \quad X = \frac{Xm\ Caw}{(1-aw)[1+(C-1)aw]} \tag{1}$$

where Xm is the moisture of the product corresponding to a monomolecular layer of absorbed water, and C is the energy constant of the product related to the heat released in the process.

2.6. Physical Characteristics

2.6.1. Color

The color of the rehydration powders (RPs) was evaluated with Color Quest XE colorimeter (Hunter Associates Laboratory, Virginia, USA) and was expressed with the L*, a*, and b* parameters.

2.6.2. Texture Profile Analysis (TPA)

The TPA of FP and RP was conducted by using a TA-XT Plus Texture Analysis (Stable Micro Systems, Godalming, UK). The factors determined were hardness (H), adhesiveness (A), cohesiveness (C), gumminess (G), and chewiness (C). The purees (50 g) were subjected to compressive force by probe, up to the distance of 5 mm, with a 1/4 "P/0-25 stainless-steel cylindrical probe. The conditions set were as follows: two penetration cycles, pre-test speed 0.5 mm/s; post-test speed 1.0 mm/s; depth of 16 mm; test time of 3 s; trigger force 5 g.

2.7. Sensory Evaluation and Acceptability

The sensory evaluation and acceptability of FP and RP were evaluated with 50 adults. The purees (10 g each) were presented in plastic cups. The acceptability was evaluated according to a hedonic scale of 9 points, with ends (1) "I do not like" and (9) "I like very much". Acceptability was calculated as the average score of the hedonic scale.

The sensory evaluation was studied by CATA (check-all-that-apply) questions. A list of 20 terms was presented to the consumers: pleasant, unpleasant, clear, dark, hard, soft, sandy, creamy, consistence, fluid, and flavor (smooth, intense, sweet, salty, cereal-like, acid, bitter, strange, artificial, rancid, and fruity). Consumers were instructed to tick the terms that most accurately describe the products. Frequency of mention for each term was determined by counting the number of consumers that used that term to describe each sample.

2.8. Statistics Analysis

All analyses were carried out in triplicate. Statistical analysis was done by using one-way analysis of variance (ANOVA). Tukey test was used to assess any differences between group means. Differences were considered significant at ϱ < 0.05. Statistic for Windows version 9.0 (USA) was used.

3. Results

3.1. Proximal Composition

Table 1 shows the proximal composition of the mixture of ingredients to elaborate the puree and DP obtained by different drying methods. The protein and ash content did not show significant variations after drying; however, lipid content significantly decreased after cooking and dehydration.

Table 1. Proximal composition of fresh puree and dry powders (g/100 g db).

Sample	Moisture	Ash	Protein	Fat	Carbohydrates
MP	9.64 ± 0.28^{b}	3.84 ± 0.20^{a}	10.71 ± 0.29^{a}	6.77 ± 0.17^{a}	78.68
LD	5.89 ± 0.19^{c}	3.82 ± 0.12^{a}	10.47 ± 0.36^{a}	5.42 ± 0.21^{b}	80.29
CD	11.04 ± 0.65^{a}	3.59 ± 0.12^{a}	10.21 ± 0.10^{a}	3.72 ± 0.16^{c}	82.47
ED	8.93 ± 0.15^{b}	3.44 ± 0.09^{a}	10.26 ± 0.35^{a}	2.66 ± 0.15^{d}	83.65

Values are means ± standard deviations from triplicate analysis. Different superscript letters in the same column indicated significant difference ($p < 0.05$). MP: mixture of ingredients to elaborate the purees; LD: lyophilized; CD: dehydrated by convection; ED: extruded.

3.2. Particle Morphology

Figure 1 shows the scanning electron micrographs of the DP. The LD had particles that were more homogeneous and smaller in size, while ED had larger particles with agglomerated appearance. The CD had particles with a collapsed appearance.

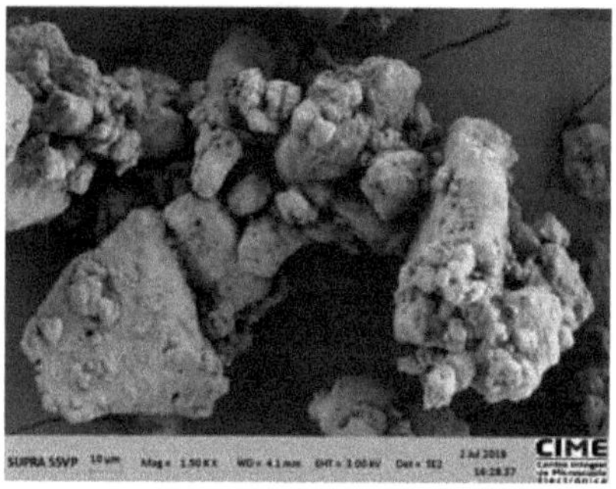

(**a**)

(**b**)

(**c**)

Figure 1. Scanning electron micrographs of dehydrated powders (1500X): (**a**) drying by forced circulation, (**b**) extruded, and (**c**) lyophilized.

3.3. Rehydration Properties

ED was the least soluble, probably due to the agglomeration of its particles [8], but it had the greatest WAC. The LD was the most soluble and had the least WAC (Table 2). The BET model had high fit with the experimental data ($R^2 > 0.90$ and %E < 3.00) (Table 2). Xm represents the optimum moisture content in which the dehydrated product will have the maximum shelf life during storage. The ED had the lowest Xm value.

Table 2. Rehydration properties of the powders obtained by different drying.

		LD	CD	ED
Solubility (g/100 g)		52.59 ± 0.33 ^{a}	48.53 ± 0.29 ^{b}	27.90 ± 0.81 ^{c}
WAC (g/g)		0.98 ± 0.27 ^{c}	1.51 ± 0.11 ^{b}	2.24 ± 0.26 ^{a}
BET model	**Xm**	0.066 ^{a}	0.062 ^{a}	0.051 ^{b}
	C	46.533 ^{a}	29.345 ^{b}	19.452 ^{c}
	R^2	0.997	0.993	0.995
	%E	1.089	2.798	2.456

Values are means ± standard deviations from triplicate analysis. Different superscript letters in the same file indicated significant difference ($p < 0.05$). LD: lyophilized; CD: dehydrated by convection; ED: extruded; Xm: moisture of the product corresponding to a saturated monomolecular layer of water; C: energy constant of the product; R^2: linear correlation coefficient; %E: relative average error percentage.

3.4. Physical Characteristics

The color and texture profile of the obtained powders and rehydrated was influenced by different drying methods (Table 3). The LD was the clearest sample (highest L*), while the powders obtained by heating (CD or ED) were darker (less L*). Hardness of FP was similar to the RP obtained by convection, because the drying was carried out at a low temperature (50 °C), avoiding the formation of hard and dry rind on the surface [4]. The rehydrated LD had less studied textural parameters compared to the other RP and FP; while the rehydrated ED had more determined textural parameters with respect to the FP and the other RP. These results agreed with those informed by Xiao et al. [4] and Wang et al. [3], respectively.

Table 3. Color and texture profile of the rehydrated powders.

		FP	LD	CD	ED
Color	L*	64.1 ± 0.4 [b]	67.4 ± 0.5 [a]	62.6 ± 0.4 [c]	59.1 ± 0.3 [d]
	a*	14.2 ± 0.4 [c]	13.1 ± 0.7 [d]	15.7 ± 0.1 [b]	19.3 ± 0.9 [a]
	b*	57.2 ± 0.4 [c]	54.8 ± 0.4 [d]	61.4 ± 0.9 [b]	64.1 ± 0.4 [a]
Texture profile	Hardness	11.5 ± 0.4 [b]	3.4 ± 0.1 [c]	10.7 ± 0.5 [b]	34.2 ± 0.7 [a]
	Adhesiveness	70.4 ± 2.6 [c]	16.8 ± 1.0 [d]	87.4 ± 4.7 [b]	212.2 ± 2.7 [a]
	Cohesiveness	0.63 ± 0.01 [d]	0.69 ± 0.01 [cd]	0.80 ± 0.01 [a]	0.67 ± 0.03 [b]
	Gumminess	7.9 ± 0.3 [b]	2.37 ± 0.08 [c]	8.5 ± 0.5 [b]	22.8 ± 0.4 [a]
	Chewiness	6.9 ± 0.8 [c]	2.2 ± 0.4 [d]	8.1 ± 0.7 [b]	21.3 ± 0.1 [a]

Values are means ± standard deviations from triplicate analysis. Different superscript letters in the same file indicated significant difference ($p < 0.05$). FP: fresh puree; LD: lyophilized; CD: dehydrated by convection; ED: extruded.

3.5. Sensory Evaluation and Acceptability

The FP was described as soft, consistent, clear, mild, pleasant, sweet, fruity, and slightly artificial taste (average acceptability of 7.4 on the hedonic scale). The PD by the different drying methods and reconstituted with water were described as follows: (i) lyophilized—soft, clear, intense flavor, bitter, strange, and unpleasant (3.7 on the hedonic scale); (ii) convection drying—soft, consistent, sandy, clear, consistent, and intense and strange taste (4.4 on the hedonic scale); and (iii) extruded—pleasant, consistent, sandy, dark, mild, fruity, and sweet (6.1 on the hedonic scale).

4. Discussion

The apparent decrease in the amount of lipids in the obtained powders was possibly due to the formation of amylose–lipid complexes during the cooking–dehydration processes [3] especially during the extrusion.

The particle size and morphology were different for each drying process, and this influenced the rehydration properties. All the DPs had pores that would facilitate hydration; however, the particles of LD showed homogenous size, which increases the exposure to water and improves the solubility. The particles of CD had contracted and collapsed appearance, which is characteristic when the water diffusion is slow and there is more time for the structures to deform. The ED presented the lowest solubility, probably due to the larger particle size and the agglomeration between them [8].

WAC was higher in this sample, possibly due to the formation of amylose–lipid complexes with water retention capacity and for the breakage of hydrogen bridges which favors hydration capacity [3]. A high WAC is a desirable feature for the preparation of soups, baby food, and instant puddings.

The RP from LP was the clearest, because during the drying by sublimation, non-enzymatic browning reactions are avoided [9]. Non-enzymatic browning reactions and sugar caramelization are promoted during high-temperature drying processes (mainly ED) [10].

The rehydrated of LD had the lowest values of hardness, adhesiveness, gumminess, and chewiness with respect to the other RD and the FP. This result coincides with that reported by [9], who explained that the ice crystals formed during freezing could break the cell structure of the sample and produce softer textures. The rehydrated ED was harder, with more adhesiveness, gumminess, and chewiness than FP and the other RP. This result could be due to the formation of amylose–lipid complexes. Moreover this result could be due to the formation of a hard crust on the particle surface by high temperatures, the rapid evaporation of water, and the high pressure generated between the particles during the process.

The rehydrated purees from LD and CD were described with negative sensory attributes (intense, bitter, and strange taste). In addition, the puree reconstituted from LD was described as being more fluid. Consumers differentiated the extruded sample by considering it darker, sweeter, and fruity, with more consistency and harder than the other samples. Therefore, the results of the evaluation of the sensory attributes by CATA method reflected the results obtained through instrumental measurements. Valentina et al. (2016) observed that, after lyophilization of some foods, some negative sensory attributes are highlighted with reduction in the acceptability of these products after lyophilization. Jafari et al. [10] observed improvement in the acceptability of bread composed of extruded sorghum-wheat, compared to that made with native sorghum-wheat. These authors explained that extrusion promotes Maillard reactions and caramelization, and therefore a darker and sweeter product is obtained.

5. Conclusions

The different drying methods influenced the technological and sensory features of the dehydrated and reconstituted powders.

Consumers were able to differentiate the samples obtained by different drying methods and attributed sensory terms to them that agreed with the instrumental determinations (such as color and texture) made in the samples.

The extruded powder had a better water retention capacity, with appropriate texture and sensory description. Therefore, the dehydrated powder that showed the best characteristics to produce a baby instant puree to reconstitute was obtained by extrusion.

Acknowledgments: This work was supported by grant IaValSe-Food-CYTED (Ref. 119RT0567), Consejo Nacional de Investigaciones Científicas y Técnicas (CONICET) and Secretaría de Ciencia y Técnica y Estudios Regionales (SECTER), Universidad Nacional de Jujuy (Argentina).

References

1. OMS; Organización mundial de la salud. *La Alimentación del Lactante y del niño Pequeño*; OPS-OMS: Washington, DC, USA, 2010; ISBN 978-92-75-33094-4.
2. Troszyŋska, A.; Szymkiewicz, A.; Wołejszo, A. The effects of germination on the sensory quality and immunoreactive properties of pea (Pisum Sativum L.) and soybean (Glycine max). *J. Food Qual.* **2007**, *30*, 1083–1100, doi:10.1111/j.1745-4557.2007.00179.x.
3. Wang, L.; Duan, W.; Zhou, S.; Qian, H.; Zhang, H.; Qi, X. Effects of extrusion conditions on the extrusion responses and the quality of brown rice pasta. *Food Chem.* **2016**, *204*, 320–325, doi:10.1016/j.foodchem.2016.02.053.
4. Xiao, M.; Yi, J.; Bi, J.; Zhao, Y.; Peng, J.; Hou, C.; Lyu, J.; Zhou, M. Modification of Cell Wall Polysaccharides during Drying Process Affects Texture Properties of Apple Chips. *J. Food Qual.* **2018**, 4510242, doi:10.1155/2018/4510242.
5. AOAC. Association of Official Analytical Chemists. Methods of Analysis (AOAC). Available online: http://www.aoac.org/ (accessed on March 2019).
6. Wani, I.A.; Sogi, D.S.; Gill, B.S. Physicochemical and functional properties of flours from three Black gram (Phaseolus mungo L.) cultivars. *Int. J. Food Sci. Technol.* **2013**, *48*, 771–777, doi:10.1111/ijfs.12025.
7. Tonon, R.; Freitas, S.S.; Hubinger, M.D. Spray drying of açai (*Euterpe oleraceae*mart.) juice: Effect of inlet air temperature and type of carrier agent. *J. Food Process. Preserv.* **2011**, *35*, 691–700, doi:10.1111/j.1745-4549.2011.00518.x.
8. Ahmed, M.; Sorifa, A.M.; Eun, J.B. Effect of pretreatments and drying temperatures on sweet potato flour. *Int. J. Food Sci. Technol.* **2010**, *45*, 726–732, doi:10.1111/j.1365-2621.2010.02191.x.
9. Valentina, V.; Pratiwi, A.R.; Hsiao, P.Y.; Tseng, H.T.; Hsieh, J.F.; Chen, C.C. Sensorial Characterization of Foods Before and After Freeze-drying. *Austin Food Sci.* **2016**, *1*, 1027–1031.
10. Jafari, M.; Koocheki, A.; Milani, E. Physicochemical and sensory properties of extruded sorghum-wheat composite bread. *J. Food Meas. Charact.* **2018**, *12*, 370–377, doi:10.1007/s11694-017-9649-4.

proceedings

Proceedings

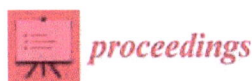

Design and Acceptability of a Multi-Ingredients Snack Bar Employing Regional PRODUCTS with High Nutritional Value †

Francisco Teodoro Rios, Argentina Angelica Amaya, Manuel Oscar Lobo and Norma Cristina Samman *

Centro de Investigaciones Interdisciplinarias en Tecnologías y Desarrollo Social para el NOA-(CIITED), CONICET-Facultad de Ingeniería, Universidad Nacional de Jujuy, Ítalo Palanca 10.4600, San Salvador de Jujuy, Jujuy, Argentina; frios8078@gmail.com (F.T.R.); argentinaamaya12@gmail.com (A.A.A.); mlobo958email2@gmail.com (M.O.L.)

* Correspondence: normasamman@gmail.com

† Presented at the 2nd International Conference of IaValSe-Food Network, Lisbon, Portugal, 21–22 October 2019.

Published: 26 August 2020

Abstract: The aim was to develop a snack bar using regional food products. The formulation included traditional cereals and amaranth, quinoa, sunflower, flax, chia, sesame and poppy seeds subjected to different treatments. Two sensory evaluations were carried out to evaluate acceptability. Snack bars containing toasted seeds presented high acceptability by the consumer. Amaranth, quinoa, chia and sunflower significantly increased the acceptability. The sensory methods applied allowed for the selection of ingredients and processing technologies that increase the preference of consumers.

Keywords: amaranth; chia; quinoa; snack bar; sensory analysis

1. Introduction

The need to have a greater number of nutritious and healthy foods leads to the development of new products, with the incorporation of ingredients or active compounds that have beneficial effects for health [1]. A balanced and healthy diet depends on many factors, among which the quality and composition of raw materials that are used in the formulation of foods stand out. Snack bars are products that allow for the incorporation of multiple ingredients that, when properly selected, can increase the nutritional and functional value of the product. Andean grains such as quinoa (*Chenopodium quinoa*) and amaranth (*Amaranthus caudatus*), recognized for the nutritional quality of their proteins and their contribution in essential fatty acids, minerals and dietary fiber, are being reintroduced in the Andean region of the Argentine Northwest [2]. On the other hand, seeds such as chia (*Salvia hispanica* L.) and sesame (*Sesamum indicum*) are ingredients that can be used in the development of snack bars. Nutritionally, the seeds stand out in general due to the high content of lipids, proteins and fibers. The lipids of chia seeds have a high content of polyunsaturated fatty acids (PUFAs), particularly omega-3 (linolenic acid) and omega-6 (linoleic acid) [3]. Therefore, they could be adequate to supplement cereals nutritionally in the formulation of snack bars. In addition to their nutritional properties, seeds provide functional components such as polyphenols and flavonoids [4]. Therefore, the seeds have significant advantages over other ingredients used in the manufacturing of bars, such as oats and rice. However, it is necessary to prepare these raw materials prior to their use, which can induce negative effects in their nutritional and functional compounds and, consequently, in the final product. An important aspect to take into account in the design and development of products is to consider the preference and acceptance of the type and quantity of any ingredient by consumers.

This work proposes to develop a snack bar with high consumer acceptability by employing nutritious regional food products rich in functional compounds.

2. Materials and Methods

2.1. Materials

Quinoa (*Chenopodium quinoa* Willd. var. Inta Hornillos) and amaranth (*Amaranthus caudatus* var. rosado) were provided by the Instituto de Investigacion y Desarrollo Tecnologico para la Agricultura Familiar (IPAF) (Jujuy-Argentina). The seeds of chia (*Salvia hispanica* L.), sesame (*Sesamum indicum*), flax (*Linum usitatissimum*), sunflower (*Helianthus annuus*) and poppy (*Papaver somniferum*) were purchased from Melar S.A. (Argentina). Puffed rice and oats (rolled) were purchased at the local market (Jujuy). In the preparation of the binder, honey obtained from producers of the region of the Yungas (Jujuy) and commercial sucrose were used. The amounts and ratio of honey/sucrose remained constant in the preparation of the bars.

2.2. Methods

The nutritional compositions of quinoa and amaranth (proteins, moisture, lipids, fiber and ash) were determined by official AOAC methods [5].

2.3. Design and Elaboration of the Snack Bars

For the elaboration of the bar, the seeds and cereals were added to the union syrup and continuously mixed until obtaining a homogeneous composition. Then, they were placed in stainless steel molds, pressed for 10 min and left to stand (1 h) at room temperature. The product was packaged and stored at room temperature up to sensory analysis [6].

The design and development of the product was carried out through a sensory analysis with 160 consumers in two stages. Consumers were recruited among students, teachers and administrative staff of the Faculty of Engineering, with an age range between 18 and 50 years, of both genders. The evaluation was carried out in the sensory analysis laboratory of the Engineering Faculty—UNJu. Each consumer was given the total samples of each experimental design in 20 g portions of each, coded by three random digits. Next to the trays, a form was provided describing the objective of the study and the instructions to carry out the analysis as well as drinking water for mouth wash between samples.

2.4. Sensory Analysis

2.4.1. Sensory Analysis 1

In the first place, the acceptability of the samples (seeds to be used as ingredients in the snack bar) was studied, to which different processes were applied depending on the temperature (T) and time (t). The applied processes were toasting (190 °C × 3 min), boiling (boiling: t1 = 15 min and t2 = 25 min), dry heating (T = 80 °C, t1 = 45 and t2 = 60 min) and baking (T = 130 °C, t = 30 and 45 min). In the first trial, a total of seven samples were analyzed (3 processes two times = 6 + 1 processes: roasting). The evaluation of acceptability was made with 75 consumers using a hedonic scale of 9 points with the following extremes: I dislike it a lot = 1 and I like it a lot = 9. In addition to the hedonic test, a survey was carried out with open questions to generate the descriptors of the samples. The attributes most frequently used (>20%) were selected as descriptive attributes of the samples.

2.4.2. Sensory Analysis 2

Once the processes of adaptation of the raw materials were defined, a second sensory analysis was carried out with 85 consumers. The acceptability of the samples was evaluated in addition to a check-all-that-apply test that included 16 attributes grouped into appearance, taste, aroma and texture. The second sensory analysis was carried out with the objective of selecting the ingredients that produce the highest preference of the samples following a Taguchi L_8 (2^7) design. According to the design, eight trials were carried out at two levels: absence and presence of each seed, which corresponds to the seven variables: quinoa, amaranth, chia, sesame, sunflower, flax and poppy.

2.5. Statistical Analysis

Samples from both trials were evaluated using the ANOVA, considering the processes and ingredients as variation factors. A principal component analysis (PCA) was applied to the descriptive data of each sensory analysis for its correlation with the acceptability of the samples.

3. Results

3.1. Sensory Analysis. Process Selection

Table 1 shows the multi-ingredient snack bar acceptability for a total of 75 consumers. Seeds and Andean grains were cooked by different processing conditions (n = 7) to study their incorporation into the bar. The ANOVA applied to the acceptability data determined significant differences ($p < 0.05$). The samples To and Ct2 presented the highest acceptability in opposition to Dt2. Figure 1 shows the PCA of the attributes and the relationship with the acceptability of the samples, which was used as a supplementary variable. The first two factors (F) explain 61.09% of the total variability of the data. The positive factor F1 indicates that the attributes that characterize the samples Ct1 and Dt1 were related to "good appearance", "bright" and "caramel color". However, the increase in acceptability was correlated to a negative F1, which is associated with samples To and Ct1, highlighting the attributes of "toasted", "crumbly in the mouth" and "very sweet". On the other hand, along the negative axis of F2, opposite to To, the samples of lower acceptance were observed: Bt1, Bt2 and Dt2. The B samples were strongly associated with the attributes "rancid taste" and "rancid smell"; for this reason, they would be rejected. In the case of D samples, although they were associated with "bright" and "good appearance", they would be rejected for "hard" and "sticky" attributes. Therefore, the processes that produced greater acceptance corresponded to the roasting of raw materials, followed by cooking.

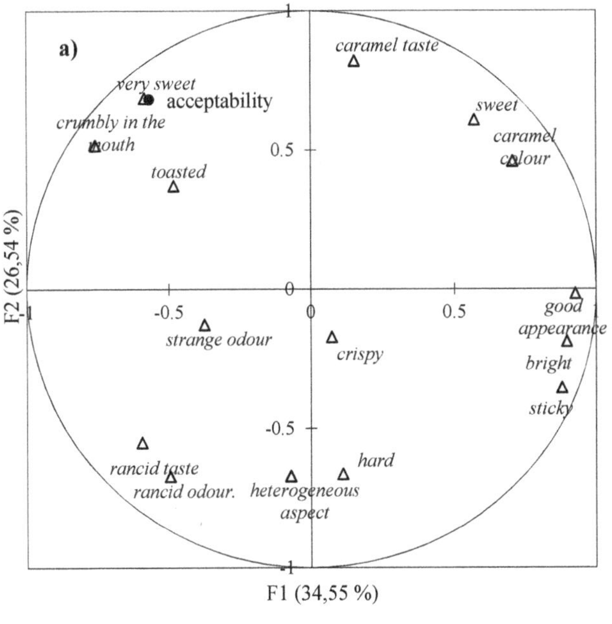

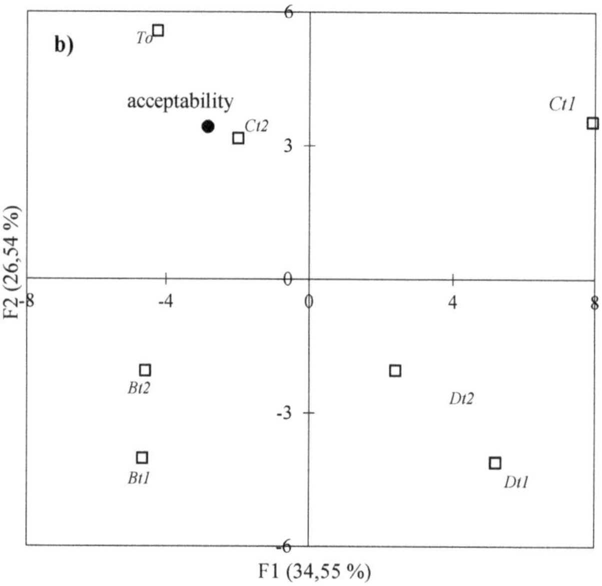

Figure 1. Representation of F1–F2 factors of the principal component analysis applied to the data of the first sensorial analysis—process selection. (**a**) Variables: sensory attributes (Δ), supplementary variable: acceptability (•). (**b**) Samples (□), supplementary variable: acceptability (•).

Table 1. Conditions and response of acceptability of the different preparation processes.

Process	Temperature (°C)	Time (min)	Acceptability (n = 75)
Toasting (To)	190	3	6.53 a
Cooking (Ct2)	94	20	6.11 a
Baking (Bt2)	130	45	5.27 b
Baking (Bt1)	130	30	5.26 b
Cooking (Ct1)	94	10	5.02 b
Dry heating (Dt1)	80	45	4.73 b
Dry heating (Dt2)	80	60	4.05 c

Different letters in the same column indicate statistically significant differences, $p < 0.05$.

3.2. Sensorial Analysis. Ingredient Selection

Table 2 shows the ANOVA applied to the samples that presented a combination of the relative content of the different ingredients for their selection and incorporation into the final product. The samples with higher proportions of amaranth, quinoa, sunflower and chia presented the highest preference. In contrast, the samples in which flax and poppy predominated (b6 and b7) were significantly ($p < 0.05$) the least accepted. In Figure 2, the PCA shows the relationship between the attributes and the acceptability of the different formulations. The first two factors explained 68.94% of the total variability. The positive F1 factor indicated a weak correlation with the increase in the acceptability of the samples. According to the F1 axis, a separation of the samples into two groups was observed.

The samples placed in the F1 positive sector were those of greater acceptance and were associated mainly with the attributes "good appearance" and "sweet taste". In the case of the negative F2 axis, there was a negative correlation ($r > 0.70$) between the acceptability and the taste and smell attributes due to rancidity. The attribute "too rancid" was perceived mainly in seeds with high lipid content, such as flax, poppy, chia and sunflower. In addition, the "heterogeneous aspect" provided by flax and poppy, related to their color and, negatively influenced the acceptability of the samples. Therefore, due to the attributes provided by the flax and poppy seeds, their incorporation into the formulation of the bars was discarded.

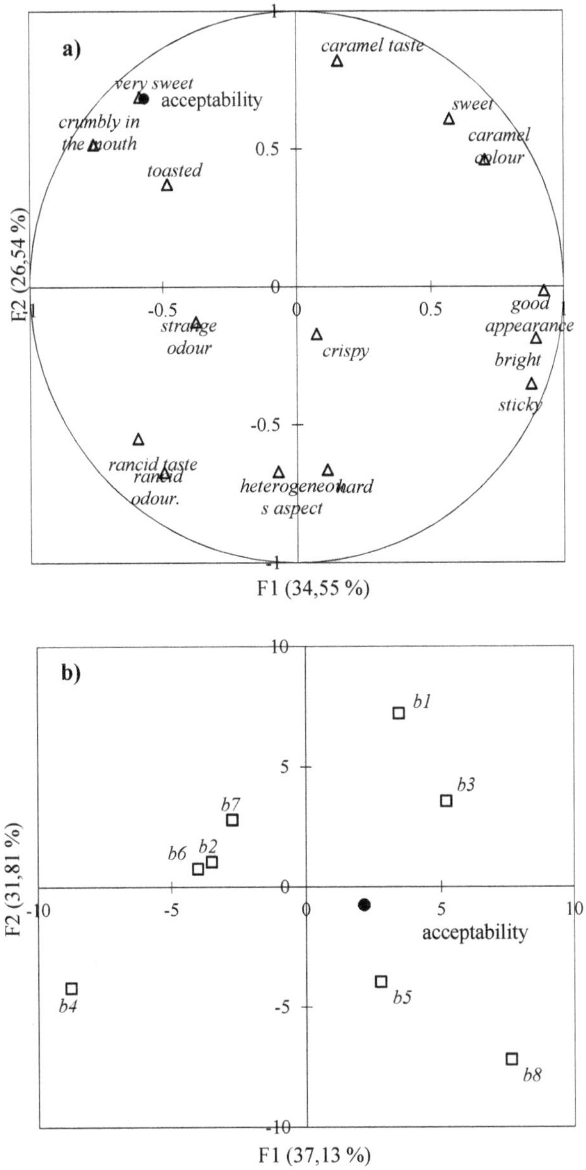

Figure 2. Representation of F1–F2 factors of the principal component analysis applied to the data of the second sensorial analysis—ingredient selection. (**a**) Variables: sensory attributes (Δ), supplementary variable: acceptability (•). (**b**) Samples (□), supplementary variable: acceptability (•).

Table 2. Variables and experimental design response to the selection of ingredients.

Sample	Amaranth	Quinoa	Chia	Sunflower	Sesame	Poppy	Flax	Acceptability (n = 85) *
b1	2	2	1	2	1	1	2	6.00 [a]
b3	2	2	1	1	2	2	1	5.82 [ab]
b2	2	1	2	2	1	2	1	5.74 [ab]
b4	**2**	**1**	2	1	2	1	2	5.70 [ab]
b5	1	2	2	2	2	1	1	5.37 [b]
b6	1	1	1	2	2	2	2	3.54 [c]
b7	1	2	2	1	1	2	2	4.42 [c]
b8	1	1	1	1	1	1	1	5.71 [ab]

* Different letters in the same column indicate statistically significant differences, $p < 0.05$.

3.3. Nutritional Analysis

Table 3 shows the nutritional composition of the raw materials and selected snack bar. The chemical composition of the seeds showed a high content of lipids, followed by proteins. On the other hand, rice and oats have a high content of carbohydrates (CH). Therefore, the bar could be correctly complemented with seeds, cereals and Andean crops in order to elaborate a product of high nutritional value and greater acceptability. In the nutritional composition of the final product, lipids and proteins were in the range of FAO's nutrient contribution recommendations. Additionally, the use of seeds in an integral way was responsible for the high content of dietary fiber.

Table 3. Proximal composition of the selected ingredients and the final product.

Ingredient	Moisture	Protein	Lipid	Ash	Fiber	Carbohydrates *
	(g/100 g)					
Oats	8.2	16.9	6.9	1.7	10.6	66.3
Sesame	3.8	17.7	49.7	4.5	18	6.4
Sunflower	3.7	19.9	56.4	2.9	2.7	17.1
Chia	7.7	20	31	4.6	31.5	5.2
Amaranth	10.5	13.4	6.4	2.9	11.3	55.3
Quinoa	11.3	12.1	6.3	2	10.4	57.2
Rice	8.6	5.9	0.7	0.5	1.7	82.5
Final product	10.2	9.7	10.5	2.3	6.5	60.8

* Calculated by difference.

4. Discussion

According to the sensory study, the most important attributes for the preferences of snack bars were the appearance, "bright" and "good appearance", followed by the taste, principally "sweet" and "toasted". The "sweet taste" attribute is one of the most important and representative of snack bars. "Sweet taste" and its intensity can mask other ingredients, like antioxidants, which usually are rejected by consumers despite the fact that their consumption may be beneficial for health [6]. However, it should be mentioned that the increase in "sweet taste" implies the addition of simple sugars, which is negative from the point of view of health recommendations.

The "rancid taste" and "rancid smell" attributes were identified as rejected attributes for different products. The attribute "rancid" is characteristic of the oxidation of the lipids of the seeds due to processing at high temperatures for long times [7]. In this sense, roasting was the better process to use due to the short time involved [8]. Despite the fact that the seed content improves the nutritional composition of the product, both the processes and the consumers did not accept a high content of the seeds in the final product. Furthermore [9] determined that the health information did not determine the selection of a snack bar. Therefore, the design of a product with high nutritional and functional value does not only depend on the health benefits but also on aspects that condition the preference of the consumers.

5. Conclusions

The sensorial methods applied allowed for the selection of ingredients and processing technologies that increase the preference of consumers, identifying attributes of acceptability and rejection.

In general, the acceptance of snack bars depended mainly on attributes such as "sweetness" and "good appearance", and rejection occurred in formulations containing ingredients with high lipid content that were treated at high temperatures for a long time.

Acknowledgments: This work was supported by grant Ia ValSe-Food-CYTED (Ref. 119RT0567) and Consejo Nacional de Investigaciones Científicas y Técnicas CONICET and Secretaria de Ciencia y Técnica y Estudios Regionales SECTER, Universidad Nacional de Jujuy (Argentina).

References

1. Sun-Waterhouse, D.; Farr, J.; Wibisono, R.; Saleh, Z. Fruit-based functional foods I: Production of food-grade apple fibre ingredients. *Int. J. Food Sci. Technol.* **2008**, *43*, 2113–2122, doi:10.1111/j.1365-2621.2008.01806.x.
2. Nascimento, A.C.; Mota, C.; Coelho, I.; Gueifão, S.; Santos, M.; Matos, A.; Sammán, N.; Castanheira, I. Characterization of nutrient profile of quinoa (*Chenopodium quinoa*), amaranth (*Amaranthus caudatus*), and purple corn (*Zea mays* L.) consumed in the North of Argentina: Proximates, minerals and trace elements. *Food Chem.* **2014**, *148*, 420–426, doi:10.1016/j.foodchem.2013.09.155.
3. Pizarro, P.L.; Almeida, E.L.; Sammán, N.C.; Chang, Y.K. Evaluation of whole chia (*Salvia hispanica* L.) flour and hydrogenated vegetable fat in pound cake. *LWT Food Sci. Technol.* **2013**, *54*, 73–79, doi:10.1016/j.lwt.2013.04.017.
4. Coelho, M.S.; de Araujo Aquino, S.; Latorres, J.M.; de las Mercedes Salas-Mellado, M. In vitro and in vivo antioxidant capacity of chia protein hydrolysates and peptides. *Food Hydrocoll.* **2019**, *91*, 19–25, doi:10.1016/j.foodhyd.2019.01.018.
5. AOAC. *Official Methods of Analysis of the Association of Official Agricultural Chemists*, 16th ed.; AOAC: Washington, DC, USA, 1995.
6. Rios, F.; Lobo, M.; Sammán, N. Acceptability of beehive products as ingredients in quinoa bars. *J. Sci. Food Agric.* **2018**, *98*, 174–182, doi:10.1002/jsfa.8452.
7. Damodaran, S.; Parkin, K.L. *Química de los Alimentos de Fennema*; Editorial Acribia: Zaragoza, Spain, 2018.
8. Repo-Carrasco-Valencia, R.A.; Encina, C.R.; Binaghi, M.J.; Greco, C.B.; Ronayne de Ferrer, P.A. Effects of roasting and boiling of quinoa, kiwicha and kañiwa on composition and availability of minerals in vitro. *J. Sci. Food Agric.* **2010**, *90*, 2068–2073, doi:10.1002/jsfa.4053.
9. Pinto, V.R.A.; de Oliveira Freitas, T.B.; de Souza Dantas, M.I.; Della Lucia, S.M.; Melo, L.F.; Minim, V.P.R.; Bressan, J. Influence of package and health-related claims on perception and sensory acceptability of snack bars. *Food Res. Int.* **2017**, *101*, 103–113, doi:10.1016/j.foodres.2017.08.062.

proceedings

MDPI

Proceedings

Development of Gluten-Free Breads Using Andean Native Grains Quinoa, Kañiwa, Kiwicha and Tarwi †

Ritva Repo-Carrasco-Valencia [1,*], Julio Vidaurre-Ruiz [1] and Genny Isabel Luna-Mercado [2]

[1] Centro de Investigación e Innovación en Productos Derivados de Cultivos Andinos CIINCA, Universidad Nacional Agraria La Molina UNALM, Avenida La Molina s/n, Lima 12, Peru; vidaurrejm@lamolina.edu.pe

[2] Facultad de Ciencias Agrarias, Escuela profesional de Ingeniería Agroindustrial, Universidad Nacional del Altiplano, Puno 21001, Peru; gluna@unap.edu.pe

* Correspondence: ritva@lamolina.edu.pe

† Presented at the 2nd International Conference of Ia ValSe-Food Network, Lisbon, Portugal, 21–22 October 2019.

Published: 26 August 2020

Abstract: The aim of this study was to develop gluten-free breads using the flours of Andean native grains. The following native grains were used: quinoa (*Chenopodium quinoa*) Pasankalla variety, kiwicha (*Amaranthus caudatus*) Centenario variety, kañiwa (*Cheopodium pallidicaule*) Illpa Inia variety and tarwi (*Lupinus mutabilis*) Blanco de Yunguyo variety. The formulations of the breads with Andean grains flours were optimized using the Mixture Design and the Central Composite Rotational Design, analyzing the dough's textural properties (firmness, consistency, cohesiveness and viscosity index), specific volume and crumb texture. Potato starch and xanthan gum were used in the preparation of the breads. The optimized formulations of the gluten-free breads with Andean grain flours were composed of quinoa (46.3%), kiwicha (40.6%), kañiwa (100%) and tarwi (12%) flours. The gluten-free breads developed showed acceptable specific volume and low crumb firmness and could help to improve the nutrition of celiac patients.

Keywords: quinoa; kiwicha; kañiwa; tarwi; gluten-free bread

1. Introduction

The gluten-free products market has increased throughout Latin America, especially in Peru, where there has been an increasing demand for gluten-free bread products in supermarkets, due to the increase in patients diagnosed with celiac disease and consumers seeking "healthier" alternatives. However, these are not necessary healthier, because the gluten-free bakery products on the market are made from starches or white rice flour, which are lacking in high quality proteins and important micronutrients.

Andean grains such as quinoa, kiwicha, kañiwa and tarwi do not contain peptides similar to wheat gluten; therefore, they are raw materials appropriate for consumption by celiacs. Likewise, these grains are sources of starches (more than 70% of their composition) [1–3], which are necessary to create bread structure [4]. On the other hand, tarwi is a legume also known as the Andean soybean because of its high protein and oil content (almost 50% and around 20%, respectively). The oil of tarwi could function as a natural emulsifier to retain the gas produced during the fermentation of gluten-free breads [5]. Quinoa is appreciated because of its high protein quality, having a balanced essential amino acid composition, and it is also considered a source of fiber and minerals [6,7]. Kiwicha is a very good source of iron, calcium and zinc. It contains more zinc and iron than conventional maize and beans [8]. Kañiwa, the least studied Andean grain, grows mainly in Peru and Bolivia, between 3500 and 4200 m above sea level where the climatic conditions are extreme [9]. Its small grains contain more protein than the common cereals; it has a good content of essential amino acids; it is rich in lysine, the first limiting amino acid in all cereals; it is rich in unsaturated fatty acids; and it is an excellent source of dietary fiber and an important source of minerals, especially iron, calcium, phosphorus and vitamins such as riboflavin [10].

The inclusion of these grains in the formulations of gluten-free breads is promising. Therefore, the aim of the research was to develop gluten-free breads with quinoa, kiwicha and kañiwa flours, and a bread with the mixture of quinoa and tarwi, using the response surface methodology, in order to find the proportions of the ingredients that can produce bread with acceptable quality.

2. Materials and Methods

2.1. Conditioning and Characterization of Raw Materials

Quinoa (*Chenopodium quinoa*) Pasankalla variety and kiwicha (*Amaranthus caudatus*) Centenario variety were supplied by the Cereals and Native Grains Program at the National Agrarian University La Molina, kañiwa (*Cheopodium pallidicaule*) Illpa Inia variety was supplied by ILLPA Puno Peru Agricultural Experimental Station, and tarwi (*Lupinus mutabilis*) Blanco de Yunguyo variety was bought from the local market of Cajamarca-Peru. Tarwi grains were conditioned, in order to eliminate the alkaloids, according to the procedure described by Jacobsen & Mujica [11]. All the grains were milled using a hammer mill (Retsch SR 300, Haan, Germany), and the proximal composition of the flours was determined following the procedures of the AOAC (2000) [12].

2.2. Experimental Design

The mixture design was used to optimize the textural properties (firmness, consistency, cohesiveness and viscosity index) of the gluten-free doughs with quinoa and kiwicha flours. For the inclusion of tarwi in the gluten-free bread formulation, the Central Composite Rotational Design (CCRD) was used; under the same principle of optimizing the dough textural properties and for the case of kañiwa bread, the CCRD was used to optimize the bread volume and backing loss.

The variables for the mixture design were the proportions of water (70–110%), xanthan gum (0.5–2%) and quinoa or kiwicha flour (10–50%). In the case of CCRD, the variables were the proportions of water (75–160%) and tarwi flour (10–30%).

In the case of the gluten-free breads with quinoa, kiwicha and tarwi, the target was to find the optimal levels of the variables that could imitate the textural properties of a dough control (3.69 ± 0.2 N of firmness; 56.5 ± 3.7 N.s of consistency; 2.7 ± 0.2 N of cohesiveness and 36.2 ± 2.1 N.s of viscosity index), which was previously developed and showed good quality properties [13].

In the case of the gluten-free bread with kañiwa, the variables were the proportions of water (75–125%), xanthan gum (0.35–0.65%) and kañiwa flour (40%). The optimization criteria were the maximization of the specific volume of the bread and the minimization of the baking loss.

2.3. Dough and Bread Preparation

For the mixture designs (doughs with quinoa and kiwicha flours), 16 formulations were used for each experiment with different proportions of water, xanthan gum, and quinoa or kiwicha flour; the other ingredients for the dough preparation were potato starch, sugar (3%), salt (2%), soybean oil (6%) and yeast (3%). For the CCRD (doughs with tarwi flour), 13 formulations with different proportions of water and tarwi flour were used; these were mixed with potato starch, quinoa flour (46%), sugar (3%), xanthan gum (0.5%), salt (2%), soybean oil (6%) and yeast (3%). For bread with kañiwa flour, the same levels of sugar, salt, soybean oil and yeast were used.

All ingredients were mixed at two different speeds for 3 min in total and used to fill aluminum molds (300 g); then, they were fermented for 30 min at 30 °C and 85–90% RH, and finally, they were baked at 200 °C for 60 min.

2.4. Dough Textural Properties

The texture analysis of the doughs (without yeast) was carried out using the Back Extrusion accessory of the INSTRON universal texturometer (Model 3365, Canton, MA, USA), where a portion of dough was deposited in the Back Extrusion cylinder (diameter, 50 mm; height, 70 mm) and penetrated up to 50% with a plunger (diameter, 42 mm) at a speed of 1 mm/s and with a trigger force of 10 gf; finally, the plunger returned to its original position at the same speed. The textural properties determined were the firmness (N), consistency (N.s), cohesiveness (N) and viscosity index (N.s).

2.5. Specific Volume

The bread volume (mL) was measured by laser topography (BVM-6610, Perten Instruments, Hägersten, Sweden), and the specific bread volume (mL/g) was calculated by dividing the volume by the bread weight.

2.6. Textural Properties of Bread

The texture profile analysis (TPA) was carried out on breads 24 h after baking using an Instron Universal Testing Machine (Model 3365, Instron Co., Canton, MA, USA), according to the procedure described by Vidaurre-Ruiz et al. (2019).

3. Results and Discussion

The proximal composition of the quinoa, kiwicha, kañiwa and tarwi flours was 8.0–11.0% of moisture, 14.0–53.0% of protein, 6.0–22.0% of fat, 2.0–3.0% of ash, 2.5–9.0% of fiber and 12.8–69.5% of carbohydrates, respectively.

The gluten-free dough optimized with quinoa flour that was able to imitate the textural properties of the dough control was composed of quinoa flour (46.3%), potato starch (53.7%), xanthan gum (0.5%) and water (75.2%). In the case of the dough with kiwicha flour, the optimized formulation consisted of kiwicha flour (40.6%), potato starch (59.4%), xanthan gum (0.5%) and water (80.9%). The dough optimized with tarwi flour was composed of tarwi flour (12%), quinoa flour (46), potato starch (42%) and water (102%). The optimized formulation of the gluten-free bread with kañiwa flour was kañiwa flour (100%), xanthana gum (0.9%) and water (140%). The rest of the ingredients remained constant, according to the type of dough, as explained above.

The quality properties of the optimized breads, such as the specific volume and crumb texture, are shown in Table 1, where it can be seen that the gluten-free breads with Andean grains had acceptable specific volumes, as well as soft crumbs (Figure 1). According to Alvarez-Jubete et al. [14], the components of Andean grains such as the fats and starches, which include low levels of amylose, significantly help to improve the quality of gluten-free breads, producing breads with soft crumbs and with a lesser tendency to retrograde, therefore increasing the shelf life of the products.

The formulation optimized with 100% kañiwa flour demonstrates that the starches of this grain are propitious for baking and that when they are mixed with the appropriate levels of water and xanthan gum, they can produce breads of good physical and nutritional quality. Likewise, tarwi flour can be included in a smaller proportion (12%) due to its high content of proteins, which have a great capacity to absorb water [5].

The inclusion of xanthan gum was minimal (0.5–0.9%) in doughs with Andean grains; this shows that the ingredients of the grains can function as natural emulsifiers; however, the use of gums is still necessary to achieve the stability of the emulsion during baking.

Table 1. Characteristics of gluten-free breads (GFB) made with Andean grains.

Quality Parameters	GFB-Quinoa	GFB-Kiwicha	GFB-Kañiwa	GFB-Tarwi
Baking loss (%)	26.9 ± 0.5	27.3 ± 0.8	28.2 ± 0.7	30.2 ± 0.8
Specific volume (mL/g)	2.3 ± 0.0	2.4 ± 0.1	2.73 ± 0.0	2.13 ± 0.00
Crumb hardness (N)	1.8 ± 0.3	3.6 ± 0.3	1.4 ± 0.2	2.3 ± 0.3
Cohesiveness	0.31 ± 0.0	0.21 ± 0.0	0.32 ± 0.00	0.39 ± 0.00
Springiness	0.8 ± 0.1	0.65 ± 0.00	0.87 ± 0.00	0.89 ± 0.00
Gumminess (N)	0.5 ± 0.1	0.8 ± 0.1	0.5 ± 0.1	0.9 ± 0.1
Chewiness (N)	0.45 ± 0.1	0.49 ± 0.0	0.38 ± 0.1	0.80 ± 0.1

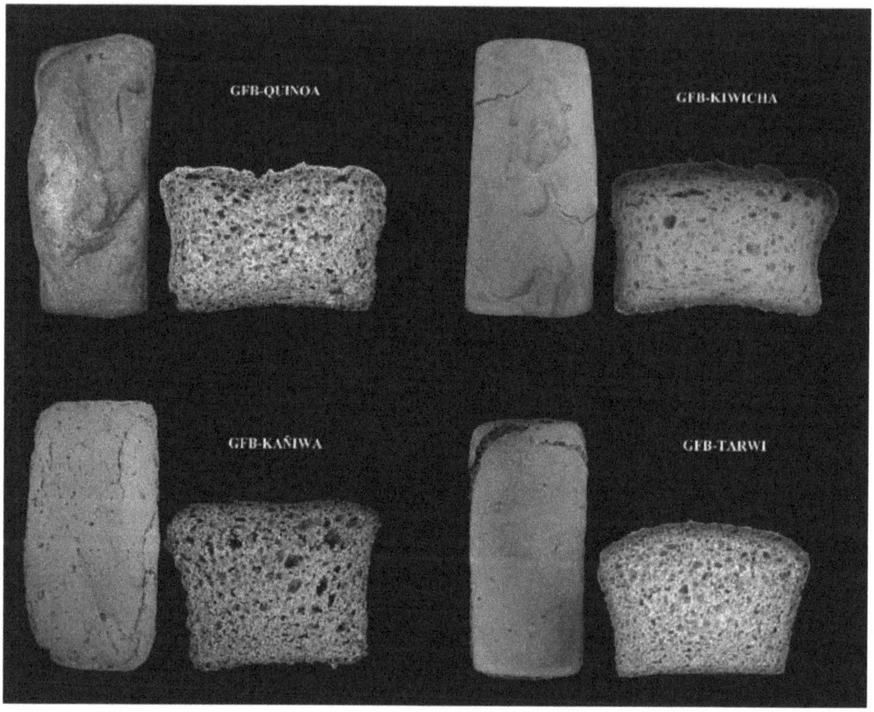

Figure 1. Representative images of gluten-free breads (GFB) with quinoa, kiwicha, kañiwa and tarwi.

4. Conclusions

It was possible to develop gluten-free breads with quinoa (46.3%), kiwicha (40.6%), kañiwa (100%) and tarwi (12%) flours. The components of the Andean grains such as the lipids and starches help to improve the quality properties of gluten-free breads, producing soft crumbs and breads with acceptable specific volumes. The inclusion of xanthan gum in the gluten-free doughs with Andean grain flours was minimal, but its use is still necessary to achieve the stability of the emulsion during the baking process. The gluten-free breads developed contained a good amount of Andean grain flours and could help improve the nutrition of celiac patients.

Funding: This work was supported by the grant Ia ValSe-Food-CYTED (Ref. 119RT0567) and PROTEIN2FOOD Project (European Union's Horizon 2020, N° 635727).

References

1. Ramos Diaz, J.M.; Kirjoranta, S.; Tenitz, S.; Penttilä, P.A.; Serimaa, R.; Lampi, A.M.; Jouppila, K. Use of Amaranth, Quinoa and Kañiwa in Extruded Corn-Based Snacks. *J. Cereal Sci.* **2013**, *58*, 59–67. doi:10.1016/j.jcs.2013.04.003.
2. Rosell, C.M.; Cortez, G.; Repo-Carrasco, R. Breadmaking Use of Andean Crops Quinoa, Kañiwa, Kiwicha, and Tarwi. *Cereal Chem.* **2009**, *86*, 386–392. doi:10.1094/CCHEM-86-4-0386.
3. Schoenlechner, R. Pseudocereals in Gluten-Free Products. In *Pseudocereals: Chemistry and Technology;* John Wiley & Sons, Ltd.: Chichester, UK, 2017; pp. 193–216, doi:10.1002/9781118938256.ch9.
4. Benavent-Gil, Y.; Rosell, C.M. *Technological and Nutritional Applications of Starches in Gluten-Free Products. Starches for Food Application*; Elsevier Inc.: London, UK, 2018. doi:10.1016/b978-0-12-809440-2.00009-5.
5. Vidaurre-Ruiz, J.M.; Salas-Valerio, W.F.; Repo-Carrasco-Valencia, R. Propiedades de Pasta y Texturales de las Mezclas de Harinas de Quinua (*Chenopodium Quinoa*), Kiwicha (*Amaranthus Caudatus*) y Tarwi (*Lupinus Mutabilis*) En un Sistema Acuoso. *Rev. Investig. Altoandinas J. High. Andean. Res.* **2019**, *21*, 5–14. doi:10.18271/ria.2019.441
6. Repo-Carrasco-Valencia, R.A.M.; Encina, C.R.; Binaghi, M.J.; Greco, C.B.; Ronayne de Ferrer, P.A. Effects of Roasting and Boiling of Quinoa, Kiwicha and Kañiwa on Composition and Availability of Minerals in Vitro. *J. Sci. Food Agric.* **2010**, *90*, 2068–2073. doi:10.1002/jsfa.4053.
7. Stikic, R.; Glamoclija, D.; Demin, M.; Vucelic-Radovic, B.; Jovanovic, Z.; Milojkovic-Opsenica, D.; Jacobsen, S.-E.; Milovanovic, M. Agronomical and Nutritional Evaluation of Quinoa Seeds (*Chenopodium Quinoa* Willd.) as an Ingredient in Bread Formulations. *J. Cereal Sci.* **2012**, *55*, 132–138. doi:10.1016/j.jcs.2011.10.010.
8. Burgos, V.E.; Binaghi, M.J.; Ronayne de Ferrer, P.A.; Armada, M. Effect of Precooking on Antinutritional Factors and Mineral Bioaccessibility in Kiwicha Grains. *J. Cereal Sci.* **2018**, *80*, 9–15. doi:10.1016/j.jcs.2017.12.014.
9. Dirección Regional Agraria Puno. *Variabilidad Genética de Cañihua En Las Provincias de Puno*; Editora DISKCOPY S.A.C.: Puno, Peru, 2012; ISBN 978-612-46286-0-3.
10. Repo-Carrasco-Valencia, R. Andean Indigenous Food Crops: Nutritional Value and Bioactive Compounds. Ph.D. Thesis, University of Turku, Turku, Finland, 2011.
11. Jacobsen, S.-E.; Mujica, A. El Tarwi (*Lupinus Mutabilis* Sweet.) y Sus Parientes Silvestres. *Bot. Econ. Los Andes Cent. Univ. Mayor San Andrés* **2006**, *28*, 458–482.
12. AOAC. *Official Methods of Analysis*, 17th ed.; Association of Official Analytical Chemists: Gaithersburg, MD, USA, 2000.
13. Vidaurre-Ruiz, J.; Matheus-Diaz, S.; Salas-Valerio, F.; Barraza-Jauregui, G.; Schoenlechner, R.; Repo-Carrasco-Valencia, R. Influence of Tara Gum and Xanthan Gum on Rheological and Textural Properties of Starch-Based Gluten-Free Dough and Bread. *Eur. Food Res. Technol.* **2019**, *245*, 1347–1355. doi:10.1007/s00217-019-03253-9.
14. Alvarez-Jubete, L.; Auty, M.; Arendt, E.K.; Gallagher, E. Baking Properties and Microstructure of Pseudocereal Flours in Gluten-Free Bread Formulations. *Eur. Food Res. Technol.* **2010**, *230*, 437–445. doi:10.1007/s00217-009-1184-z.

proceedings

Proceedings

Nutritional Composition and Uses of Chia (*Salvia hispanica*) in Guatemala †

Elsa Julieta Salazar de Ariza *, Ana Ruth Belloso Archila, Ingrid Odete Sanabria Solchaga and Sandra Beatriz Morales Pérez

Department of Chemical and Pharmaceutical, School of Nutrition, Universidad de San Carlos de Guatemala, Cdad. de Guatemala 01012, Guatemala; anabelloso@gmail.com (A.R.B.A.); odette_2865@hotmail.com (I.O.S.S.); sanbmor@gmail.com (S.B.M.P.)

* Correspondence: julietasalazar10@gmail.com

† Presented at the 2nd International Conference of Ia ValSe-Food Network, Lisbon, Portugal, 21–22 October 2019.

Published: 10 September 2020

Abstract: *Salvia hispanica* L. (chia) is a seed native to Mexico and Central America; in Guatemala it is known as "chan". It is cultivated in small quantities and sold in neighborhood markets in different areas of the country. Little information exists on the nutritional composition of this seed, so chia samples were obtained in five regions of the country and studied for their macronutrients, minerals, and fatty acids, the form of consumption, and the adequate quantity to mix with water. We found an average of 22% for protein, 18.6% for fat, 67% for alpha-linoleic acid, 19% for raw fiber, and 9 mg/100 g of sodium, among other nutrients. The most frequent form of consumption is mixed with lemonade; the primary known benefits are to lose weight, improve digestion, and as a source of fiber; the adequate amount to mix with water is 0.8% of gel equivalent to 0.4% of seeds.

Keywords: alfa-linoleic acid; chia; consumers; *Salvia hispanica* L.

1. Introduction

Salvia hispanica L. is a seed native to Mexico and Central America; it has been known and cultivated in Guatemala since pre-Hispanic time [1,2], its common name is "chia", and traditionally, it has been sold in bulk in neighborhood markets, even though recently it has also been sold packaged and branded in supermarkets and gourmet stores. It is also found as an ingredient in processed food, which allows it to be advertised as a functional food. The chia seed has a high content of α-linoleic acid, protein, and dietary fiber, propertied which make it recommended for consumption as flour, as oil, or as a hydrolyzed protein [3].

Alvarado [4] quantified the macronutrients and fatty acids of the chia seed cultivated in the northern region of Guatemala, finding amounts of these nutrients similar to those reported by Bushway et al. [5], Ayerza and Coates [6] and Ixtaina [7].

The objective of the present study was to quantify the macronutrients, minerals, and fatty acids of *S. hispanica* sold in bulk in Guatemala, as well as to determine the form of consumption, the consumer benefits known by consumers, and the concentration of the gel that they consider suitable when mixed with water.

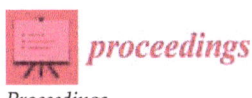

2. Materials and Methods

2.1. Sample Collection

The chia seeds were obtained in bulk in the markets of Cobán, Alta Verapaz (northern region), Atiquizaya, El Salvador (southern region), Jocotán, Chiquimula (eastern region), San Juan Ostuncalco (western region), Chimaltenango and San Agustín Acasaguastlán, El Progreso (central region).

2.2. Macronutrient, Minerals and Fatty Acids Quantification

Proximal analysis was used to determine the macronutrients. The humidity was determined in a convection oven at 60 °C, until it was a constant weight; the raw protein by the Kjeldhal method using Kjeltec Auto 1030 Analyzer equipment; the raw fiber by acid and alkaline digestion using Fibertec System I equipment; the fat by petroleum benzene extraction using Goldfish equipment, carbohydrates by difference, and energy by the Atwater factor. The minerals were determined by atomic absorption spectroscopy, using Perkin Elmer AAnalyst 100 equipment, except phosphorus, which was determined using a UV/VIS Lambda 11 colorimeter. The fatty acids were quantified using a GCMS-QP2020 gas chromatograph, with a 30-m-long DB-5MS column, an internal diameter of 0.25 mm, and a stationary phase thickness of 0.2 micrometers. The column temperature in the oven was 100.0 °C.

2.3. Preparation of Chia Gel Solutions

The gel was prepared by hydrating the chia seeds in water, at a 5:95 proportion, letting them stand for 24 h, and then separating the water and the gel by simple filtration. With the gel obtained, solutions were prepared at 0.4, 0.6, 0.8, and 1.6% in water, which were identified by three-digit codes, obtained from a random number table.

2.4. Sensory Study

We chose 50 young adults, students, and workers from the Universidad de San Carlos de Guatemala, who self-identified as chia consumers. In an appropriate room, they were presented with an ounce of each solution of gel in water, they were invited to taste each one and to qualify them as "lacking", "sufficient", or "a lot", which were respectively recorded as 1, 2, and 3; afterwards, they were asked two open questions: in what form have they consumed chia and what health benefits does it have. They were chia consumers.

2.5. Statistical Analysis

Macronutrient, mineral, and fatty acid content data were analyzed with descriptive statistics; the information about forms of consumption was analyzed by means of frequencies, and variance analysis was applied to the rating given to the chia gel solutions, through the Microsoft Excel 2010 statistical analysis.

3. Results and Discussion

3.1. Macronutrients, Minerals, and Fatty Acidsin S. hispanica L.

Table 1 shows the average content and standard deviation of macronutrients, minerals, and fatty acids of the chia samples collected from neighborhood markets from five regions of the country.

The chia is confirmed as a seed with high protein content, similar to the protein content of the bean (*Phaseolus vulgaris*), 20 to 22%, but with a digestibility of 29%, which is considerably lower than that of the bean, which is 79% [3,8]. In this study, 3% more protein was found than that reported by Alvarado [4], which is an acceptable variation, since values have been reported in the 18 to 24 g/100 g range [3,9].

The high content of alpha-linoleic acid in chia was also confirmed, representing 66% of total fatty acid of that oil seed, a similar concentration to that reported by Alvarado [4] and by Ixtaina [7].

As for fat, it is notable that a much lower value was found than the range reported by the aforementioned authors, which is between 29 and 34 g/100 g.

The mineral content is similar to the values found in the United States Department of Agriculture database [10], which highlights its low sodium content and its high content of calcium, potassium, and magnesium.

Table 1. Macronutrient, energy, mineral, and fatty acid content in *S. hispanica* L.

Nutrient	Average (n = 5)	SD
Macronutrients, g/100 g		
Water	5	0.7
Protein	22.1	0.7
Fat	18.7	2.7
Carbohydrates	31.8	2.7
Raw fiber	19	1.1
Energy, Kcal/100g	383	17
Minerals, mg/100 g		
Calcium	512	50
Phosphorus	156	27
Potassium	722	164
Magnesium	358	24
Sodium	9	1
Iron	5	2
Manganese	2.5	1.1
Copper	0.9	0.4
Zinc	4.7	0.5
Fatty acids, g/100 g		
α-linoleicacid	66.8	7.5
Linoleic acid	17.2	1.4
Palmitic acid	8.7	3.1
Stearic acid	4.8	1.1
Oleic acid	1.1	0.2

3.2. Forms of Consumption and Benefits of Consumption of *S. hispanica*

Table 2 shows the forms of chia consumption in lemonade and water. It is important to observe that all the forms of consumption refer to the raw, whole seed, which has a digestibility of 29%, which means that only 6.38% of the protein and 19% of the alpha-linoleic acid are used. This indicates the need to promote the production of chia flour and its use in different preparations, through which up to 80% digestibility can be reached [3]. Despite the aforementioned, consumption of raw chia seeds is still important due to their raw fiber and dietary fiber content, made of xylose, glucose, and glucoronic acid, which increases the feeling of satiety and thereby decreases energy consumption, and decreases obesity, cardiovascular diseases, and type 2 diabetes risk factors [9,11].

The following benefits were mentioned regarding the consumption of chia seeds: it helps weight loss, improves digestion as a fiber source, prevents constipation as a source of omega-3 fatty acids, controls blood lipids and glycaemia. This indicates that consumers have correct information on the benefits of whole raw chia seed consumption, with the exception of the benefit of being an omega-3 fatty acid source. On the other hand, although it was requested that they mention only the benefits of consumption, three of the 50 consumers spontaneously mentioned that there is a risk of appendicitis, due the possibility of seeds accumulation on it.

3.3. Adequate Quantity to Add to Water

Table 3 shows the rating given by the consumers about the solutions of water and chia seed gel. The consumers' opinion did not show significant differences ($p \geq 0.05$) for concentrations between 0.4 and 0.6%, indicating that they are lacking; the opinions were significantly different ($p \leq 0.05$) when rating solution with 0.8 and 1.6%, indicating that they were sufficient and a lot, respectively.

Table 2. Form of consumption and benefits mentioned by consumers of *S. hispanica* L. in Guatemala.

Form of Consumption	Frequency
In lemonade	33
Mixed with water	15
In prepared drinks or juices	14
In cookies	7
In smoothies	7
Mixed with yogurt	5

Table 3. Percentage of *S. hispanica* gel adequate for a drink according to consumers' tastes.

Percentage of Gel	Rating *
0.4	1.48
0.6	1.58
0.8	1.98
1.6	2.62

* 1 = lacking, 2 = sufficient, 3 = a lot.

Young Guatemalan adults, as residents of a tropical country, are in the habit of consuming water frequently throughout the day, so they carry a container of about a liter of water with them, to which they could add chia seeds. Taking into account that the chia seed doubles its volume due to gel formation when put in contact with water and left to stand for 2 h, 4 g of chia seeds could be added to a liter of water and consumed over the course of the day, which would allow the ingestion of 0.76 g of raw fiber and 1.26 g of dietary fiber, as well as 0.25 g of usable protein and 0.15 g of usable alpha-linoleic acid. The quantities are relatively small in relation to the recommended daily intake of nutrients, hence the need to promote chia consumption in the form of flour and through other preparations, in order to increase daily intake.

4. Conclusions

Chia (*Salvia hispanica* L.) has 22% protein, 18% fat, 31% carbohydrates, 19% raw fiber, and 67% alpha-linoleic acid, as well as 9 mg/100 g of sodium, 512 mg/100 g of calcium, 722 mg/100 g of potassium and 358 mg/100 g of magnesium. In Guatemala, it is consumed as a raw seed mixed with lemonade and the benefits it provides as a source of dietary fiber are recognized. The consumer considers it adequate to use 0.8% gel mixed with water, which is equivalent to 4 g of chia seeds.

Funding: This work was supported by grant IaValSe-Food-CYTED (119RT0567) and Universidad de San Carlos de Guatemala.

References

1. Ullah, R.; Nadeem, M.; Khalique, A.; Imran, M.; Mehmood, S.; Javid, A.; Hussain, J. Nutritional and therapeutic perspectives of Chia (*Salvia hispanica* L.): A review. *J. Food Sci. Tech.* **2015**, *53*, 1750–1758, doi:10.1007/s13197-015-1967-0.
2. Di Sapio, O.; Bueno, M.; Busilacchi, H.; Quiroga, M.; Severin, C. Caracterización Morfoanatómica de Hoja, Tallo, Fruto y Semilla de *Salvia hispanica* L. (Lamiaceae). *Boletín Latinoamericano y del Caribe de Plantas Medicinales y Aromáticas* **2012**, *11*, 249–268.
3. Monroy-Torres, R.; Mancilla-Escobar, M.L.; Gallaga-Solórzano, J.C.; Medina-Godoy, S.; Santiago-García, E.J. Protein Digestibility of chia seed (*Salvia hispanica* L). *Revista de Salud Pública y Nutrición* **2008**, *9*, 1–9.

4. Alvarado Rupflin, D.I. Caracterización de la semilla del chan (*Salvia hispanica* L.) y diseño de un producto funcional que la contiene como ingrediente. *Revista de la Universidad del Valle de Guatemala* **2011**, *23*, 43–49.
5. Bushway, A.; Belyea, P.R.; Bushway, R.J. Chia seed as a source of oil, polysaccharide and protein. *J. Food Sci.* **1981**, *46*, 1349–1356, doi:10.1111/j.1365-2621.1981.tb04171.
6. Ayerza, R.; Coates, W. Influence of environment on growing period and yield, protein, oil and alpha-linoleic content of three chia (*Salvia hispanica* L.) selections. *Ind. Crops Prod.* **2009**, *30*, 321–324, doi:10.1016/j.indcrop.2009.03.009.
7. Ixtaina, V.Y.; Martínez, M.L.; Spotorno, V.; Mateo, C.M.; Maestri, D.M.; Diehl, B.W.K.; Nolasco, S.M.; Tomás, M. Characterization of chia seed oil obtained by pressing and solvent extraction. *J. Food Comp. Anal.* **2011**, *24*, 166–174, doi:10.1016/j.jfca.2010.08.006.
8. Cárdenas Quintana, H.; Gómez Bravo, C.; Díaz Novoa, J.; Camarena Mayta, F. Evaluación de la calidad de la proteína de 4 variedades mejoradas de frijol. *Rev. Cuba. Aliment Nutr.* **2000**, *14*, 22–27.
9. Olivos-Lugo, B. L.; Valdivia-López, M. Á.; Tecante, A. Thermal and physicochemical properties and nutritional value of the protein fraction of Mexican chia seed (*Salvia hispanica* L.). *Food Sci. Technol. Int.* **2010**, *16*, 89–96. doi.org/10.1177/1082013209353087.
10. USDA-Database. Available online: https://ndb.nal.usda.gov./ndb/search/list (accessed on 19 May 2019).
11. Coelho, M.S.; Salas-Mellado, M.M. Revisão: Composição química, propriedades funcionais e aplicações tecnológicas da Chia (*Salviahispânica* L.) em alimentos. *Braz. J. Food Technol.* **2015**, *17*, 259, doi:10.1590/1981-6723.1814.

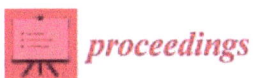

proceedings

Proceedings

Chia (*Salvia hispanica*): Nutraceutical Properties and Therapeutic Applications †

Talía Hernández-Pérez [1], María Elena Valverde [1], Domancar Orona-Tamayo [2] and Octavio Paredes-Lopez [1,*]

[1] Centro de Investigación y de Estudios Avanzados del IPN, Departamento de Biotecnología y Bioquímica, Km. 9.6 Libramiento Norte Carretera Irapuato-León, Irapuato, Guanajuato CP 36824, Mexico; talia.hernandez@cinvestav.mx (T.H.-P.); malevalverde@gmail.com (M.E.V.)

[2] Centro de Innovación Aplicada en Tecnologías Competitivas, Omega No. 201 Col. Industrial Delta, León, Guanajuato CP 37545, Mexico; domancar@gmail.com

* Correspondence: octavio.paredes@cinvestav.mx; Tel.: +52-(462)-6239674

† Presented at the 2nd International Conference of Ia ValSe-Food Network, Lisbon, Portugal, 21–22 October 2019.

Published: 1 September 2020

Abstract: Chia seeds (*Salvia hispanica* L.) have high amounts of nutraceutical compounds and a great commercial potential. The aim of this work was to identify proximate composition, fatty acids profile, total phenolics and antioxidant capacity of chia, as well the protein fractions and determine their antihypertensive potential. The seeds exhibited high content of protein, fiber, and lipids, mainly polyunsaturated fatty acids. Important amounts of phenols and a high antioxidant activity (DPPH and ABTS) were found. Globulins fraction showed the most abundant concentration followed by albumins. Peptides from albumins and globulins exhibited the strongest potential against the angiotensin-converting enzyme (ACE) activity. In brief, this study demonstrates that chia can be considered a seed with high nutritional content, antioxidant activity and as a novel antihypertensive agent; important factors for the frequent incorporation of chia in the human diet.

Keywords: antioxidants; antihipertensive; chia; phenolics; nutraceutical; *Salvia hispanica*

---

1. Introduction

Salvia genus has around 900 species and belongs to the Plantae Kingdom and Lamiaceae family. Chia (*Salvia hispanica* L.) is an annual herb that can grow up to 1 m tall and has oppositely arranged leaves with small white or purple flowers, and oval seeds showing black, gray, and black spotted to white color (Figure 1) [1,2].

Chia is native to central Mexico up to northern Guatemala and began to be used as food in 3500 BC. Chia was a main crop of pre-Columbian societies; Aztecs, Mayas, and Incas (1500–900 BC) used it as medicine, food, painting, and as energizer. Aztecs received it as annual tributes, and as offering to gods. The Spanish conquest suppressed its use due to religious beliefs. In 2009, chia was approved as novel food by the European Parliament and the European Council. Chia seeds are good source of macronutrients, B vitamins, calcium, phosphorus, potassium, magnesium, iron, zinc, and copper, as well as phenolic compounds. Additionally, it can be incorporated to celiac diets due to the absence of gluten [2,3]. Nowadays, chia has shown several health benefits, i.e., antioxidant potential and antihypertensive, between others [1,4]. Thus, the aim of this work was to determine the proximate composition, fatty acid profile, total phenolic content and antioxidant capacity of chia, as well as its protein fractions in terms of their antihypertensive potential.

Figure 1. Chia plants, flowers and seeds. (**a**) Purple and white flowers, (**b**) Seeds from different commercial lines cultivated in Mexico.

2. Materials and Methods

Seeds from four varieties of commercial chia (*Salvia hispanica*): Black from Puebla, White and Pinta Jalisco (PJ) from Jalisco, and Xonotli from Guanajuato, Mexico, were soaked in distilled water (1:10, *w*/*v*) for 1 h to allow mucilage production. They were frozen overnight (−80 °C) and freeze-dried, the dry mucilage was removed mechanically. Mucilage-free seeds were milled into flour and passed through a 0.5 mm mesh to obtain a uniform particle size. The flour was defatted with hexane (1:10, *w*/*v*) in a Soxhlet unit at 65–70 °C and dried overnight at room temperature to remove remaining hexane; a second grinding was performed to obtain a smaller particle size (0.18 mm), and it was stored at 4 °C until use. Proximate composition was evaluated based on Sandoval-Oliveros and Paredes-López [5] method. Fatty acid profile was made using Guzmán-Maldonado et al. [6] procedure. Total phenolic compounds and antioxidant capacity were performed according to Martínez-Cruz and Paredes-López [3]. Antihypertensive potential was determined using the method reported by Orona-Tamayo et al. [7].

3. Results

3.1. Proximate Composition

The different types of chia analyzed presented an average of 22.7, 32.5 and 33.6 g/100 g dry basis of protein, lipids and dietary fiber, respectively (Table 1). As can be observed, these results agree with those previously reported by Ayerza and Coates [1]. Soluble and insoluble fiber fractions found in chia samples may improve glucose and lipidic profile, and intestinal function, thus reducing the risk for obesity, coronary heart disease, type II diabetes mellitus, metabolic syndrome and several types of cancer. It also is associated to increase post-meal satiety, decreasing subsequent hunger [6]. Chia fiber content (35–40 g/100 g) is equivalent to 100% of the daily recommendations for adult population [8]. On the other hand, protein content in chia seeds was higher than most of the traditionally staple grains, i.e., wheat (14%), corn (14%), rice (8.5%), oats (15.3%), and barley (9.2%) [1].

Table 1. Proximate composition and dietary fiber of chia seed.

Component	g/100 g Dry Basis
Moisture	4.5 ± 0.0
Lipids	32.5 ± 2.7
Protein	22.7 ± 0.7
Ash	3.7 ± 0.3
Soluble Dietary fiber	8.2 ± 0.8
Insoluble Dietary fiber	25.4 ± 2.2
Total Dietary fiber	33.5 ± 2.7
Carbohydrates (by difference)	3.1

Values are the mean ± SD of three determinations.

3.2. Fatty Acids Profile

S. hispanica seeds are worldwide recognized for their high content of lipid, which comprised mainly polyunsaturated fatty acids that play an important role in health [5,9]. The average amount of fatty acids in the chia varieties evaluated was 11.9 and 87.6 g/100 g of saturated and polyunsaturated fatty acids (PUFAs), respectively (Table 2). As it can be seen, three healthy fatty acids were identified in elevated amounts in the chia samples: linolenic (ω-3), linoleic (ω-6), and oleic (ω-9). Otherwise, the saturated fatty acids, palmitic (16:0) and stearic (18:0), were found in very low concentrations. These data are in accordance with previous studies by Ayerza and Coates [9] and Mohd Ali et al. [2]. They have established that *S. hispanica* is an outstanding source of PUFAs (25–40%), comprising 55–60% linolenic and 18–20% linoleic acids. Chia oil has the highest percentage (of any plant source) of *α*-linolenic acid, which is considered essential because the human body cannot produce it and is also a potent lipid antioxidant [9,10].

Table 2. Fatty acid profile of chia cultivated in Mexico.

Fatty Acid		g/100 g Oil
Polyunsaturated	Linolenic (ω-3)	59.1 ± 0.23
	Linoleic (ω-6)	21.2 ± 0.12
Monounsaturated	Oleic (ω-9)	7.3 ± 0.09
	Total	87.6
Saturated	Palmitic (16:00)	7.9 ± 0.15
	Estearic (18:00)	4.0 ± 0.06
	Total	11.9

Values are the mean ± SD of three determinations.

3.3. Protein

The mean content of protein (22.7 g/100 g dry basis) of the evaluated chia seeds was high and is similar to that reported by Ayerza and Coates [1], and it provides essential amino acids to the daily diet [5]. Several important storage protein fractions have been found within chia. Based on solubility criteria, storage protein fractions were extracted. Table 3 shows the concentration of protein fractions from the four chia varieties evaluated. The fraction of globulins was the most abundant in all the lines studied, it ranged from 11.6 to 15.5 μg/mL, followed by the fraction of albumins 9.5 to 13.1 μg/mL, and glutelins with a concentration of 7.4 to 8.7 μg/mL, and the fraction of prolamins was found in the lowest concentration 4.4 to 5.2 μg/mL. These data were comparable to results obtained by Sandoval-Oliveros and Paredes-López [5] and Orona-Tamayo et al. [7].

Table 3. Concentration of proteic fractions from chia seeds.

Line	Albumins	Globulins	Prolamins	Glutelins
	(μg/mL, Dry Basis)			
Xonotli	10.8 ± 0.9	14.6 ± 0.1	4.4 ± 0.1	8.1 ± 0.3
Pinta Jalisco	9.5 ± 0.3	11.6 ± 0.0	4.4 ± 0.1	8.3 ± 0.1
Black	13.1 ± 0.5	15.5 ± 0.1	5.2 ± 0.1	7.4 ± 0.1
White	10.8 ± 0.3	13.2 ± 1.2	4.8 ± 0.0	8.7 ± 0.2

Values are the mean ± SD of three determinations.

3.4. *Phenolic Compounds and Antioxidant Capacity*

The concentration of total phenolic compounds of the *S. hispanica* lines evaluated was in the range of 0.78 and 0.97 g/100 g dry basis (Table 4). It can be observed that chia comprises a high concentration of phenolic compounds, thus, the antioxidant capacity was evaluated using the radical scavenging assays, DPPH (2,2-diphenyl-1-picrylhydrazyl) and ABTS [2,2′-azinobis (3-ethylbenzothiazoline-6-sulphonic acid)]. Values for DPPH were in the range between 1.0 and 1.2 μg/g, and for ABTS from 1.0 to 1.6 μg/g. The concentration of total phenols and the antioxidant capacity of the chia seeds cultivated in different States of Mexico are remarkable and consistent with results obtained by Martínez-Cruz and Paredes-López [3] from *S. hispanica* cultivated in the Central area of Mexico.

Table 4. Total phenolic compounds and antioxidant capacity in chia.

Line	Total phenols (g/100 g)	DPPH (IC_{50}, μg/g)	ABTS (IC_{50}, μg/g)
Xonotli	0.78 ± 0.04	1.0 ± 0.05	1.6 ± 0.2
Pinta Jalisco	0.92 ± 0.17	1.1 ± 0.1	1.4 ± 0.03
Black	0.97 ± 0.03	1.1 ± 0.07	1.0 ± 0.1
White	0.83 ± 0.05	1.2 ± 0.02	1.4 ± 0.005

DPPH, 2,2-diphenyl-1-picrylhydrazyl; ABTS, 2,2′-azinobis (3-ethylbenzothiazoline-6-sulphonic acid).
Values are the mean ± SD of three determinations.

3.5. *Antihypertensive Effect*

Chia contains high concentration of hydrophobic amino acids (proline, leucine, phenylalanine, and isoleucine) that generate peptides with high angiotensin converting enzyme (ACE)-inhibitory activity [11]. ACE controls blood pressure by regulating the volume of fluids in the body. It converts the hormone angiotensin I to the active vasoconstrictor angiotensin II. Our results indicate that the peptides obtained from the fraction of globulins promoted the highest inhibitory effect against ACE with an IC_{50} of 203.61, 148.23 and 110.11 μg/mL for White, Black, and Pinta Jalisco chia lines, respectively. Albumins, glutelins and prolamins demonstrated lower capacity against ACE, which is in accordance to data reported by Orona-Tamayo et al. [7]. They found that peptides from albumin and globulin fractions exhibited the highest ACE-inhibitory activity (IC_{50} 377 and 339 μg/mL, respectively), followed by chia seed flour (IC_{50} 516 μg/mL). It is interesting to note that albumin and globulins from Adzuki or red beans (*Vigna angularis*) [12] required higher protein concentration to inhibit ACE enzyme activity than the equivalent proteins of chia seeds from our study.

4. Conclusions

Salvia hispanica seeds showed important contents of proteins, dietary fiber and healthy lipids. In addition, they were a good source of total phenols and have high antioxidant capacity, which suggest that the scavenge ROS (reactive oxygen species) and may decrease or prevent chronic degenerative diseases, cancer and ageing. The chia protein profile showed that globulins are the major protein fraction followed by albumins, glutelins and prolamins. Chia peptides from the protein fractions exhibited capacity as ACE inhibitors, particularly, peptides from globulin and albumin fractions showed the strongest potential against ACE. These results can suggest that chia peptides could be outstanding natural antioxidants and ACE inhibitors for human health. Further research is required to develop novel chia cultivars with better nutraceutical attributes.

Acknowledgments: This work was supported by grant Ia ValSe-Food-CYTED (119RT0567) and Consejo Nacional de Ciencia y Tecnología (Conacyt), Mexico.

References

1. Ayerza, R.; Coates, W. Ground chia seed and chia oil effects on plasma lipids and fatty acids in the rat. *Nutr. Res.* **2005**, *25*, 995–1003, doi:10.1016/j.nutres.2005.09.013.
2. Mohd Ali, N.; Yeap, S.K.; Ho, W.Y.; Beh, B.K.; Tan, S.W.; Tan, S.G. The promising future of chia, *Salvia hispanica* L. *J. Biomed. Biotechnol.* **2012**, *2012*, 171956, doi:10.1155/2012/171956.
3. Martínez-Cruz, O.; Paredes-López, O. Phytochemical profile and nutraceutical potential of chia seeds (*Salvia hispanica* L.) by ultrahigh performance liquid chromatography. *J. Chromatogr. A* **2014**, *1346*, 43–48, doi:10.1016/j.chroma.2014.04.0079.
4. Muñoz, L.A.; Cobos, A.; Diaz, O.; Aguilera, J.M. Chia seed (*Salvia hispanica*): An ancient grain and a new functional food. *Food Rev. Int.* **2013**, *29*, 394–408, doi:10.1080/87559129.2013.818014.
5. Sandoval-Oliveros, M.R.; Paredes-López, O. Isolation and characterization of proteins from chia seeds (*Salvia hispanica* L.). *J. Agric. Food Chem.* **2013**, *61*, 193–201, doi:10.1021/jf3034978.
6. Guzmán-Maldonado, S.H.; Vazquez, M.G.; Aguirre, J.A.; Serrano, F.I. Contenido de ácidos grasos, compuestos fenólicos y calidad industrial de maíces nativos de Guanajuato. *Rev. Fitotec. Mex.* **2015**, *38*, 213–222, ISSN 0187-7380.
7. Orona-Tamayo, D.; Valverde, M.E.; Nieto-Rendón, B.; Paredes-López, O. Inhibitory activity of chia (*Salvia hispanica* L.) protein fractions against angiotensin I-converting enzyme and antioxidant capacity. *LWT Food Sci. Technol.* **2015**, *64*, 236–242, doi:10.1016/j.lwt.2015.05.033.
8. Kaczmarczyk, M.M.; Miller, M.J.; Freund, G.G. The health benefits of dietary fiber: Beyond the usual suspects of type 2 diabetes mellitus, cardiovascular disease and colon cancer. *Metabolism* **2012**, *61*, 1058–1066, doi:10.1016/j.metabol.2012.01.017.
9. Ayerza, R.; Coates, W. Protein content, oil content and fatty acid profiles as potential criteria to determine the origin of commercially grown chia (*Salvia hispanica* L.). *Ind. Crops Prod.* **2011**, *34*, 1366–1371, doi:10.1016/j.indcrop.2010.12.007.
10. Marineli, R.S.; Lenquiste, S.A.; Moraes, E.A.; Maróstica, M.R. Antioxidant potential of dietary chia seed and oil (*Salvia hispanica* L.) in diet-induced obese rats. *Food Res. Int.* **2015**, *76*, 666–674, doi:10.1016/j.foodres.2015.07.039.
11. Segura-Campos, M.R.; Peralta-González, F.; Chel-Guerrero, L.; Betancur-Ancona, D. Angiotensin I-converting enzyme inhibitory peptides of chia (*Salvia hispanica*) produced by enzymatic hydrolysis. *Int. J. Food Sci.* **2013**, *8*, 158482, doi:10.1155/2013/158482.
12. Durak, A.; Baraniak, B.; Jakubczyk, A.; Świeca, M. Biologically active peptides obtained by enzymatic hydrolysis of Adzuki bean seeds. *Food Chem.* **2013**, *141*, 2177–2183, doi:10.1016/j.foodchem.2013.05.012.

proceedings

MDPI

Proceedings

Effect of Chia Seed Oil (*Salvia hispanica* L.) on Cell Viability in Breast Cancer Cell MCF-7 †

Armando M. Martín Ortega and Maira Rubí Segura Campos *

Facultad de Ingeniería Química, Universidad Autónoma de Yucatán, Periférico Norte Km. 33.5, Tablaje Catastral 13615, Col. Chuburná de Hidalgo Inn, Mérida, Yucatán C.P. 97203, Mexico; armandomarorte@gmail.com

* Correspondence: maira.segura@correo.uady.mx

† Presented at the 2nd International Conference of Ia ValSe-Food Network, Lisbon, Portugal, 21–22 October 2019.

Published: 1 September 2020

Abstract: Worldwide, cancer represents one of the main causes of mortality and morbidity, with breast cancer being the most diagnosed and the main cause of mortality among women. The purpose of this study is to evaluate the effect of chia seed oil on cell viability in the breast cancer line MCF-7. Tumor cells were treated to various concentrations of chia seed oil (12.5–400 μg/mL), then cellular viability was evaluated by (3-(4,5-dimethyl thiazole-2yl)-2,5-diphenyl tetrazolium bromide) MTT assay. Cellular viability was increased in the highest concentration group. Chia seed oil in high concentrations could potentially increase the viability of breast cancer cells.

Keywords: alternative; cancer; chia; nutraceutical; nutrition; oil

1. Introduction

Cancer is one of the main public health problems worldwide and represents the third leading cause of death in Mexico. Moreover, its incidence is increasing, without discriminating against countries or regions [1,2]. At the same time, national and international efforts to find more effective and less harmful treatments have been increased, including the investigation of bioactive compounds derived from food [3,4].

In addition to being an energy and structural source for the human body, fatty acids are bioactive lipids that regulate a large number of cellular processes, including growth regulation, apoptosis and cell proliferation [5,6]. Several in vitro and in vivo studies with isolated fatty acids and food oils have shown to have both anti-cancer and carcinogenic effects [7–10]. Thus, demonstrating an important role in the prevention and treatment of cancer. For its part, chia seed is an important source of ω-3 polyunsaturated fatty acids, and is considered a functional food because of its ability to exert anti-inflammatory, lipid-lowering, anti-hyperglycemic and metabolic regulating effects that are important in the treatment of chronic diseases, mainly metabolic [11,12]. Moreover, at present, there are international collaborations (Chia-Link International Network) for continuous research on the potential health benefits of chia seed, and the development of the functional foods derived from it.

The study of the effect of chia seed oil on cancer cells, would allow us to know its anti-cancer or carcinogenic potential, which would provide a basis for further studies in in vivo and clinical models. Likewise, the results of this study would serve as a guide for conducting studies of other oils, increasing the knowledge of the relationship between food and cancer, ultimately for the improvement of nutritional interventions in this disease.

2. Materials and Methods

2.1. Seed Oil Extraction

Dried whole seeds without mucilage were pressed in a cold pressing system until the oil was extracted, using a pressure of 8 ton/m^2. The extracted oil was stored in an amber glass container inside a refrigerator at 4 °C away from the light, to allow sedimentation of seed residues and subsequent removal by centrifugation.

2.2. Chemical Oil Hydrolysis (Ethanolysis)

To obtain the free fatty acids (FFA), *Salvia hispanica* L. oil was chemically hydrolyzed by an alkaline hydrolysis with KOH and ethanol, following the methodology proposed by Riss et al. (2013). Then, 25 g of oil was mixed with 150 mL of 1 M KOH (95% EtOH) and placed in a 65 °C water bath for 2 h. Subsequently, to stop the hydrolysis, 100 mL of distilled water was added to the mixture. The non-hydrolyzed portion was removed by extraction with 100 mL of hexane. The aqueous alcohol phase, which contained the FFA, was acidified to pH 1 with 6N HCl to remove K from the carboxyl groups of fatty acids (R-COOK + HCl → R-COOH + KCl). The resulting free fatty acids were recovered by extraction with hexane and distilled water was added until a neutral pH was obtained. The phases formed by the mixture of water and hexane were separated with a separating funnel. Finally, the upper portion, which contains the FFA, was dried with anhydrous magnesium sulfate and the solvent was evaporated with a broken steam under vacuum at 35 °C [13].

2.3. Determination of the Fatty Acid Profile by Gas Chromatography

The composition of fatty acids was determined using the methodology proposed by Us-Medina (2015) with some modifications. First, 50 mg of oil was taken in a 50 mL test tube, 10 mL of 10% *w/v* KOH in a methanol solution was added and allowed to reflux for 45 min in a controlled temperature bath (60 °C). At the end of the saponification, the sample was washed three times with 3 mL of hexane. Next, 2 mL of concentrated HCl was added and the fatty acids were extracted with three 2 mL portions of hexane, followed by drying with a nitrogen flow. Subsequently, a transesterification of the sample was performed by adding 420 μL of 5% HCl in a methanol solution to the fatty acids, then refluxing at 85 °C for 150 min in a controlled bath. The result of the transesterification was methyl fatty acid esters (FAME), which were extracted with three 2 mL portions of hexane. Then, 80 μL of a 10,000-ppm solution of the C17 standard in hexane was added. The hexenic phase was dried by a stream with nitrogen. After drying, the FAME were reconstituted with 1 mL of hexane to be injected into the chromatograph in split 25:1 mode with Helium (analytical grade) as the mobile phase. The gas chromatograph that was used is an Agilent Technologies 6890N, with an SP-2560 column, 100 m long, 0.25 mm internal diameter and 0.20 μm thick. The conditions that were used are: injector temperature of 250 °C, column flow of 1.0, oven temperature of 140 °C for 5 min and increased to 240 °C in a gradient of 4 °C/min, with a mass detector.

2.4. Evaluation of Cytotoxic and Antiproliferative Activity In Vitro

The cytotoxic and antiproliferative activities of the hydrolyzed oil were carried out by culturing the cell line with different concentrations of chia seed oil leaving them with the treatment for 48 h. The negative control was the cell line cultured only with the culture medium, while for the positive control Taxol was used, a drug commonly used in cancer chemotherapy and in vitro studies. At the end of the treatment, the MTT assay (3-(4,5-dimethylthiazol-2-yl)-2,5-diphenyl-tetrazolium bromide) was performed to evaluate cell viability (explained below).

2.4.1. Preparation of the Compounds

The hydrolyzed oil was prepared at different concentrations for evaluation. A stock of 4 mg/mL of each compound was prepared by diluting 20 mg of the compound in 5 mL of fresh medium and subsequently passed through a 0.2 μm pore size nylon membrane syringe filter (cat. 431224, Corning,

Monterrey, NL, Mexico). Serial dilutions were being made from the stock with the culture medium until concentrations of 12.5, 25, 50, 100, 200 and 400 μg/mL. These dilutions were prepared immediately before use.

2.4.2. Cell Culture

Cells were grown in Dulbecco's Modified Eagle Medium (DMEM/F-12) medium without phenol red (cat. D2906, Sigma Aldrich, St. Louis, MO, USA) supplemented with 1.2 g/L $NaHCO_3$ (cat. S5761, Sigma Aldrich, St. Louis, MO, USA) and 10% phosphate buffered saline (PBS) (cat. 10437028, Invitrogen, Carlsbad, CA, USA). The cells were incubated at 37 °C with 5% CO_2 and a humidified atmosphere, in a Lab-Line incubator (Barnstead, Melrose Park, IL, USA). Each time the cell culture reached about 70–80% confluence, subcultures were performed.

2.4.3. Cell Count and Viability

The cell count was performed with the Neubauer chamber and the determination of cell viability by the trypan blue exclusion technique.

For this procedure, a 200 μL sample of the suspended cells was taken, 20 μL of trypan blue was added, placed in a Neubauer chamber and observed under a microscope. Dead cells have a blue color. A concentration of cells was satisfied per μL, counting the living cells.

2.4.4. Evaluation of Cytoxicity and Antiproliferative Effect by the MTT Technique

The determination of cellular cytoxicity/antiproliferative was performed using the MTT technique. This technique allows the proliferation to be measured indirectly by the detection of the coloration caused by the metabolic reduction in the tetrazolium salt bromide of 3-(4,5-dimethylthiazol-2-yl)-2,5-diphenyltetrazolium (MTT), yellow, due to the action of mitochondrial dehydrogenase enzymes, so only viable cells can reduce it. The resulting compound, formazan blue, can be solubilized and quantified spectrophotometrically [14].

2.4.5. Procedure

The procedure performed is described below.

1. For each cell line, the cells were shown to have viability greater than 95% and were inoculated in 96-well plates (cat. 83.1835, Sarstedt, Newton, NC, USA). To each well was added 100 μL of prepared cell suspension at 1×10^5 cell/mL in DMEM/F-12 medium. Some wells were inoculated with culture medium only for control. The plate edge wells were inoculated with 1XPBS to prevent an evaporation of the samples.
2. After inoculation, the cells were incubated at 37 °C with 5% CO_2 for 24 h.
3. After 24 h, the medium was removed and 100 μL of the dilutions of the compounds per well (triplicate) were added. The final concentrations in the wells were: 12.5, 25, 50, 100, 200 and 400 μg/mL. Then, 100 μL of the culture medium was added to the controls of the cells without compounds and the medium without cells. The plates were incubated for 48 h.
4. At the end of the incubation period, the wells of the plates were visualized under a microscope to verify that there was no visible contamination.
5. A wash with 1X PBS was performed on all wells and left with 100 μL of 1X PBS.
6. Subsequently, 10 μL of the MTT reagent (5 mg/mL, cat. M5655 Sigma Aldrich, St. Louis, MO, USA) was added to each well and incubated for 3 h.
7. Next, 100 μL of isopropanol/DMSO (1:1) was added to each well and vigorously resuspended to solubilize the formazan crystals.
8. Finally, the plates were read on a microplate spectrophotometer at a wavelength of 590 nm.
9. The percentage of cell proliferation (% P) was calculated, with respect to the control, with the following formula:

$$\% P = (ODw/compound)/(OD\ control) \times 100$$

where:

ODw/compound: optical density of cells with compound.

OD control: optical density of the cell control (cells without compounds).

2.5. Statistical Analysis

All results were processed using descriptive statistics using measures of central tendency (mean) and dispersion (standard deviation). The data obtained from the biological activities were evaluated by means of one-way analysis of variance and a comparison of means (Student's T method) to establish statistical differences between treatments with a 95% confidence level ($p < 0.05$).

3. Results and Discussion

3.1. Gas Chromatography

The results of the gas chromatography demonstrated the presence of palmitic acid (7.7 ± 0.19%), stearic (4.45 ± 0.15%), oleic (9.57 ± 0.27%), arachidonic (0.30 ± 0.00%), linoleic (20.56 ± 0.14%) and alpha-linolenic (57.29 ± 0.07%). The fatty acid present in a greater proportion was alpha-linoleic acid, which is considered an essential fatty acid, and omega 3. The content of the latter is lower compared to that reported from other crops of Mexican origin [15]. However, it is known that the chemical composition of the seed is variable and depends on the region where it grows, with elevation above sea level being a determining factor.

3.2. Cell Viability Assay

Cellular viability was 159.8 ± 4.5% (p = 0.01), 144.4 ± 3.6% (p = 0.01), 109.5 ± 2.9% (p = 0.06), 105.1 (p = 0.09), 97.4 ± 3.1% (p = 0.31) and 91.8 ± 5.1% (p = 0.04) for 400, 200, 100, 50, 25 and 12.5 μg/mL of chia oil, respectively, compared with cells without treatment (control group: culture medium) (Figure 1). Cellular viability was significantly increased in the two major concentrations of the oil (400 and 200 μg/mL) and reduced in the lower concentration of the oil (12.5 μg/mL).

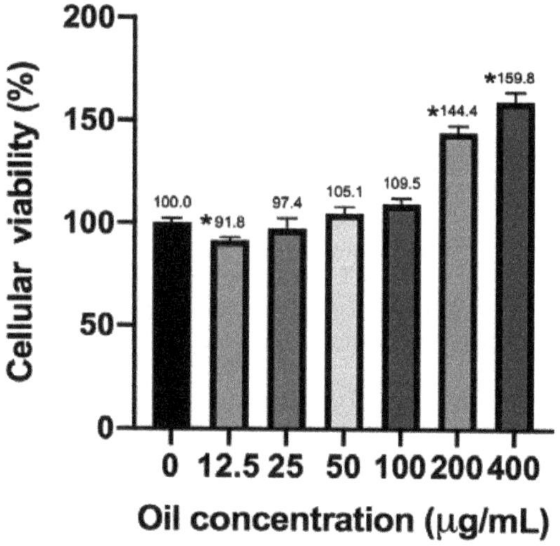

* Statistically significant

Figure 1. Cellular viability assay results.

The lower concentration (12.5 μg/mL) significantly reduced cell viability compared to the control. However, concentrations greater than 50 μg/mL increased cell viability. The two highest concentrations significantly increased cell viability. These results are due to the fact that alpha linolenic acid (omega 3) has antitumor potential, and at low concentrations (≤25 μg/mL) of chia oil its effect was prevalent. However, at high concentrations (>25 μg/mL) of the oil, the concentrations of linoleic acid (omega 6) are increased, which has been shown to increase the cell proliferation of breast cancer cells. In turn, it is important to consider that different studies have found an antitumor effect of alpha linolenic acid in various cell lines, including breast, colon and prostate cancer [16–18], while linoleic acid has demonstrated inverse effects, increasing the proliferation of breast cancer cell lines [19–21].

It is important to consider that the effects demonstrated in the *in vitro* study of the oil are not extrapolated at the systemic level, this is due to the possible metabolism and absorption of fatty acids by other organs before reaching the breast tissue. On the other hand, Espada et al. (2007) evaluated the effect of *Salvia hispanica* and *Carthamus tinctorius* oil on eicosanoid production, growth and metastasis in a murine model of mammary gland adenocarcinoma. This study found that the diet with chia oil produced a reduction in the amount of arachidonic acid and eicosanoids in the neoplasic cells ($p < 0.05$), as well as in the weight and number of tumor metastases ($p < 0.05$), compared with the *Carthamus tinctorius* diet and control diet. In addition, animals fed with chia oil showed a greater infiltration of T lymphocytes and apoptosis of the tumor cells with respect to the other diets ($p < 0.05$). Thus, the study authors concluded that *Salvia hispanica* oil is a rich source of polyunsaturated fatty acids ω-3 with the potential to inhibit tumor growth and metastasis, at least in the murine model [8].

Interestingly, other studies conducted with seed oils such as flaxseed, canola, walnut, squash and neem, have found growth-inhibiting effects of various cancer cell lines [22–25]. In addition to the potential regulatory effect of fatty acids in oils, it is possible that the presence of phytochemicals may be contributing to the antitumor effects. However, in the case of the present study, it is likely that the linoleic acid content of chia seed oil has been the main contributor to the increase in cell viability at higher concentrations.

4. Conclusions

This study suggests that chia seed oil in high concentrations could potentially increase the viability of breast cancer cells. However, at low concentrations it could reduce cell viability. Thus, future research is necessary, specifically as regards employing the isolates of omega 6 and omega 3 to extend beyond our concluded results.

Funding: This work was supported by grant Ia ValSe-Food-CYTED (119RT0567).

References

1. Stewart, B.W.; Wild, C.P. *World Cancer Report 2014*; World Health Organization: Geneva, Switzerland, 2014, doi:9283204298.
2. INEGI. *Estadísticas a Propósito del día Mundial Contra el Cáncer (Comunicado de Prensa Number 61/18)*; Instituto Nacional de Estadística y Geografía: Aguascalientes City, Aguascalientes, Mexico, 2017. Available online: https://www.inegi.org.mx/contenidos/saladeprensa/aproposito/2018/cancer2018_nal.pdf (accessed on 1 September 2019).
3. Hudson, T.J.; Anderson, W.; Aretz, A.; Barker, A.D.; Bell, C.; Bernabé, R.R., Wainwright, B.J. International network of cancer genome projects. *Nature* **2010**, *464*, 993–998, doi:10.1038/nature08987.
4. Pratheeshkumar, P.; Son, Y.O.; Korangath, P.; Manu, K.; Siveen, K. Phytochemicals in cancer prevention and therapy. *BioMed Res. Int.* **2015**, *8*, 324021, doi:10.1155/2015/324021.
5. Michalak, A.; Mosińska, P.; Fichna, J. Polyunsaturated fatty acids and their derivatives: Therapeutic value for inflammatory, functional gastrointestinal disorders, and colorectal cancer. *Front. Pharmacol.* **2016**, *7*, 459, doi:10.3389/fphar.2016.00459.
6. Gu, Z.; Suburu, J.; Chen, H.; Chen, Y.Q. Mechanisms of omega-3 polyunsaturated fatty acids in prostate cancer prevention. *BioMed Res. Int.* **2013**, *2013*, 824563, doi:10.1155/2013/824563.
7. Corsetto, P.A.; Montorfano, G.; Zava, S.; Jovenitti, I.E.; Cremona, A.; Rizzo, A.M. Effects of n-3 PUFAs on breast cancer cells through their incorporation in plasma membrane. *Lipids Health Dis.* **2011**, *10*, 73, doi:10.1186/1476-511X-10-73.

8. Espada; C.E.; Berra, M.A.; Martinez, M.J.; Eynard, A.R.; Pasqualini, M.E. Effect of Chia oil (*Salvia hispanica*) rich in ω-3 fatty acids on the eicosanoid release, apoptosis and T-lymphocyte tumor infiltration in a murine mammary gland adenocarcinoma. *Prostaglandins Leukot. Essent. Fat. Acids* **2007**, *77*, 21–28, doi:10.1016/j.plefa.2007.05.005.
9. Schley, P.D.; Jijon, H.B.; Robinson, L.E.; Field, C.J. Mechanisms of omega-3 fatty acid-induced growth inhibition in MDA-MB-231 human breast cancer cells. *Breast Cancer Res. Treat.* **2005**, *92*, 187–195, doi:10.1007/s10549-005-2415-z.
10. Song, K.-S.; Jing, K.; Kim, J.-S.; Yun, E.-J.; Shin, S.; Seo, K.-S., Park, J.-H.; Heo, J.-Y.; Kang, J.X.; Suh, K.-S.; et al. Omega-3-Polyunsaturated Fatty Acids Suppress Pancreatic Cancer Cell Growth in vitro and in vivo via Downregulation of Wnt/Beta-Catenin Signaling. *Pancreatology* **2011**, *11*, 574–584, doi:10.1159/000334468.
11. Segura-Campos, M.R.; Ciau-Solís, N.; Rosado-Rubio, G.; Chel-Guerrero, L.; Betancur-Ancona, D. Chemical and functional properties of chia seed (*Salvia hispanica* L.) gum. *Int. J. Food Sci.* **2014**, doi:10.1155/2014/241053.
12. Ullah, R.; Nadeem, M.; Khalique, A.; Imran, M.; Mehmood, S.; Javid, A.; Hussain, J. Nutritional and therapeutic perspectives of Chia (*Salvia hispanica* L.): A review. *J. Food Sci. Technol.* **2016**, *53*, 1750–1758, doi:10.1007/s13197-015-1967-0.
13. Riss, T.L.; Moravec, R.A.; Niles, A.L.; Duellman, S.; Benink, H.A.; Worzella, T.J.; Minor, L. Cell Viability Assays. *Assay Guid. Man.* **2013**, *114*, 785–796, doi:10.1016/j.acthis.2012.01.006.
14. Mosmann, T. Rapid colorimetric assay for cellular growth and survival: Application to proliferation and cytotoxicity assays. *J. Immunol. Methods* **1983**, *65*, 55–63, doi:10.1016/0022-1759(83)90303-4.
15. Porras-Loaiza, P.; Jiménez-Munguía, M.T.; Sosa-Morales, M.E.; Palou, E.; López-Malo, A. Physical properties, chemical characterization and fatty acid composition of Mexican chia (*Salvia hispanica* L.) seeds. *Int. J. Food Sci. Technol.* **2014**, *49*, 571–577, doi:10.1111/ijfs.12339.
16. Bratton, B.A.; Maly, I.V.; Hofmann, W.A. Effect of polyunsaturated fatty acids on proliferation and survival of prostate cancer cells. *PLoS ONE* **2019**, *14*, e0219822, doi:10.1371/journal.pone.0219822.
17. Chamberland, J.P.; Moon, H. Down-regulation of malignant potential by alpha linolenic acid in human and mouse colon cancer cells. *Fam. Cancer* **2014**, *14*, 25–30, doi:10.1007/s10689-014-9762-z.
18. Wiggins, A.K.A.; Kharotia, S.; Mason, J.K.; Thompson, L.U. α -Linolenic Acid Reduces Growth of Both Triple Negative and Luminal Breast Cancer Cells in High and Low Estrogen Environments. *Nutr. Cancer* **2015**, *67*, 1001–1009, doi:10.1080/01635581.2015.1053496.
19. Connolly, J.M.; Liu, X.H.; Rose, D.P. Dietary linoleic acid-stimulated human breast cancer cell growth and metastasis in nude mice and their suppression by indomethacin, a cyclooxygenase inhibitor. *Nutr. Cancer* **1996**, *25*, 231–240, doi:10.1080/01635589609514447.
20. Reyes, N.; Reyes, I.; Tiwari, R.; Geliebter, J. Effect of linoleic acid on proliferation and gene expression in the breast cancer cell line T47D. *Cancer Lett.* **2004**, *209*, 25–35, doi:10.1016/j.canlet.2003.12.010.
21. Rodriguez-Monterrosas, C.; Diaz-aragon, R.; Cortes-Reynosa, P. Linoleic acid induces an increased response to insulin in MDA-MB-231 breast cancer cells. *J. Cell Biol.* **2018**, *119*, 5413–5425, doi:10.1002/jcb.26694.
22. Sharma, R.; Kaushik, S.; Shyam, H.; Agarwal, S.; Kumar, A. Neem Seed Oil Induces Apoptosis in MCF-7 and MDA MB-231 Human Breast Cancer Cells. *Asian Pac. J. Cancer Prev.* **2017**, *18*, 2135–2140, doi:10.22034/APJCP.2017.18.8.2135.
23. Hardman, W.E.; Ion, G. Suppression of Implanted MDA-MB 231 Human Breast Ca.ncer Growth in Nude Mice by Dietary Walnut. *Nutr. Cancer* **2008**, *60*, 666–674, doi:01635580802065302.
24. Cho, K.; Mabasa, L.; Park, C.S. Canola Oil Inhibits Breast Cancer Cell Growth in Cultures and in vivo and Acts Synergistically with Chemotherapeutic Drugs. *Lipids* **2010**, *45*, 777–784, doi:10.1007/s11745-010-3462-8.
25. Truan, J.S.; Chen, J.; Thompson, L.U. Flaxseed oil reduces the growth of human breast tumors (MCF-7) at high levels of circulating estrogen. *Mol. Nutr. Food Res.* **2010**, *54*, 1414–1421, doi:10.1002/mnfr.200900521.

proceedings

MDPI

Proceedings

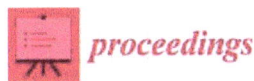

Spray-Air Contact in Tall and Short-Type Spray Dryers Affects Important Physicochemical Properties of Microencapsulated Chia Seed Oil †

María Gabriela Bordón [1,2], Noelia P. X. Alasino [3], Maria Victoria Defaín Tesoriero [4], Nahuel Camacho [5], Maria C. Penci [1,2,3] and Marcela L. Martínez [1,3,6] and Pablo D. Ribotta [1,2,3,*]

1 Instituto de Ciencia y Tecnología de los Alimentos, Facultad de Ciencias Exactas, Físicas y Naturales (ICTA-FCEFyN)–Universidad Nacional de Córdoba (UNC), 5000 Córdoba, Argentina; gabrielabordon90@gmail.com (M.G.B.); cpenci@gmail.com (M.C.P.); marcelamartinez78@hotmail.com (M.L.M.)
2 Instituto de Ciencia y Tecnología de Alimentos Córdoba (ICYTAC, CONICET-UNC), 5000 Córdoba, Argentina
3 Departamento de Química Industrial y Aplicada (FCEFyN–UNC), 5000 Córdoba, Argentina; nalasino@gmail.com
4 Grupo de Sistemas de Liberación Controlada, Centro de Química, Instituto Nacional de Tecnología Industrial (INTI), B6000XAL Buenos Aires, Argentina; mdefain@inti.gob.ar
5 Unidad de Investigación y Desarrollo en Tecnología Farmacéutica (UNITEFA, CONICET-UNC), 5000 Córdoba, Argentina; nahuelc03@gmail.com
6 Instituto Multidisciplinario de Biología Vegetal (IMBIV, CONICET-UNC), 5000 Córdoba, Argentina
* Correspondence: pribotta@agro.unc.edu.ar
† Presented at the 2nd International Conference of Ia ValSe-Food Network, Lisbon, Portugal, 21–22 October 2019.

Published: 27 August 2020

Abstract: The aim of this work was to determine the influence of the spray dryer's aspect ratio (height/diameter) on the physico-chemical properties of microencapsulated chia oil (CSO). Two different dryers were analyzed: a tall-type dryer [H/D = 5/1], and a short-type dryer [H/D = 1.65/1]. The former corresponded to a co-current flow, while the latter had a central air disperser in the chamber, and a rotary air flow. Emulsions were prepared by homogenization of CSO, and a mixture of soy protein isolate (SPI) and gum arabic (GA). The co-current contact in the tall-type dryer yielded greater oxidative stability indexes (OSI) (three times higher than CSO), which was possibly associated the reduced thermal degradation. It can be concluded that a co-current contact constitutes a better alternative for the protection of CSO.

Keywords: chia seed oil; dryer aspect ratio; microencapsulation; oxidative stability; physical properties; spray–air contact

1. Introduction

Chia seed (*Salvia hispanica* L.) oil (CSO) is the major vegetable source of alpha-linolenic acid (ALA, C18:3); thus, the development of food fortified with chia has been extensively encouraged [1] However, CSO is highly susceptible to oxidation due to the unsaturated structure of its fatty acids [2], which ultimately decreases the nutritional value of the foods and negatively impacts their sensory properties [1]. Hence, the need for the protection of Omega-3-rich oils, through microencapsulation technologies is justified [1].

Spray drying is the most widespread technology in the microencapsulation field due to its low cost, flexibility and scalability [2]. Rapid water removal results in increased product shelf-life, and reduced shipping and storage costs [3,4]. The manner in which the sprayed feed contacts the drying air has a great bearing on the dried product's properties, due to its influence on the droplet behavior during drying. This contact is dictated by the position of the atomizer in relation to the air disperser [4]. Therefore, the spray can be directed into hot air with a co-current, counter-current or mixed flow [3]. Co-current configurations are preferred if heat-sensitive products (like CSO) are involved, given that the spray evaporation is rapid, the drying air cools accordingly, and the evaporation times are short. Nonetheless, any areas of back mixing in the drying chamber (local counter-current flow) may create an overly hot environment for heat-sensitive ingredients [4]. This important issue, although very clear, is rarely addressed in the scientific literature, with only theoretical studies [4].

Based on the above considerations, the present work aimed to analyze the influence of two different spray dryers' aspect ratios (height/diameter), and the spray–air contact therein, on the physico-chemical properties of microencapsulated CSO.

2. Materials and Methods

2.1. Materials

Chia seed oil (CSO) was extracted from seeds coming from Salta province (Nutracéutica Sturla SRL, Argentina), according to Martínez et al. [5], in a pilot plant screw press (Komet Model CA 59 G, IBG Monforts, Mönchengladbach, Germany). Soy protein isolate (SPI) SUPRO E with 90% protein on fat-free basis was purchased from The Solae Company (San Isidro, Argentina); gum arabic (GA) (Alland & Robert, Paris and Normandy, France) and maltodextrin (MD) DE 5 (Lorelite 5, Companhia Lorenz, Indaial, Brazil) were purchased from a local distributor (Distribuidora NICCO, Córdoba, Argentina).

2.2. Emulsion Preparation and Characterization

Coarse emulsions were prepared by high-speed homogenization of CSO, and a mixture of SPI and GA (15,000 rpm, 2 min, Ultraturrax homogenizer IKA T18, Janke & Kunkel GmbH, Staufen, Germany); 1/1 SPI/GA and 2/1 ((SPI+GA)/CSO) ratios (w/w) were used. The coarse emulsions were further homogenized in a high-pressure valve homogenizer at 700 bar (1 cycle, EmulsiFlex C5, Avestin, Ottawa, ON, Canada). The pH of fine emulsions was adjusted to 3.0 to induce coacervation between SPI and GA, and the reaction was completed with stirring at room temperature. Finally, MD DE 5 as filler was incorporated before drying to achieve a 22% w/v final total solid content.

Particle size distribution of final emulsions was determined according to Us-Medina et al. (2018), and with a LA 950V2 Horiba (Kyoto, Japan) analyzer.

Time-dependent steady shear properties of emulsions were evaluated using a controlled-stress rheometer MCR 301 Anton Paar, equipped with a plate-cone geometry (50 mm diameter) and working with a 0.05-mm gap [6].

The morphology of oil droplets was assessed with a confocal scanning laser microscope (Olympus FV1000, Tokyo, Japan) according to González et al. [1], with brief modifications. The continuous phase was labeled with a fluorescent marker, Rhodamine B (Sigma-Aldrich, Darmstadt, Germany) (0.08 g kg^{-1} on a dry matter basis).

2.3. Spray Drying Experimental Design

Two replicated factorial designs were carried out in different spray dryers. The factors analyzed were inlet air temperature (3 levels) and feed flow rate (2 levels). The first design was performed in a tall-form spray dryer (TF-SD), Büchi B-290 (Büchi Labortechnik AG, Flawil, Switzerland), equipped with a two-fluid nozzle atomizer. The aspect ratio was 5/1 (height/diameter = 0.55 m/0.11 m). The evaluated values of inlet air temperature and feed flow rate were 130, 160 and 190 °C, and 2.6 and 5.8 mL/min, respectively. Finally, the second design was performed in a short-form spray dryer (SF-SD), Niro Mobile Minor (Søborg, Denmark), with a rotary atomizer. The aspect ratio was 1.65/1 (height/diameter = 1.40 m/0.85 m). The evaluated inlet air temperatures were the same as the previous design, and the feed flow rates were 10 and 15 mL/min.

2.4. Powder Characterization

The powders obtained with both spray dryers were characterized in terms of moisture content (MC), water activity (a_w) and particle size distribution (d_{43} and d_{32} mean diameters, and d_{50} refer to volume-based distribution) as described by Us-Medina et al. (2018). The aggregation index (AI) of microparticles after the drying of chia oil-in-water emulsions was calculated according to Ma et al. [7].

Color determinations were performed with a CM600d spectrophotometer (Konica-Minolta, Tokyo, Japan) according to González et al. [1]. Whiteness (WI) and yellowness (YI) index, as well as the color change (ΔE index) between a blank and an oil-loaded microparticle, were estimated as described by Rodriguez et al. [8]. Surface fat (SF) and encapsulation efficiency (EE) were assessed according to González et al. [1]. The powder flowability was determined using the Carr's Index (CI) and the Hausner Ratio (HR), as described by Rodriguez et al. [8]. The oxidative stability index (OSI) was determined by the Rancimat test (100 °C, 20 L/h air flow rate) according to González et al. [1]. The thermal behavior of powders was evaluated by thermogravimetric analysis (TGA), and was compared with a blank microparticle. Samples were heated from 25 to 350 °C with a linear rate of 10 °C/min. Glass transition temperatures were determined by differential scanning calorimetry (DSC, TA Instruments, New Castle, DE, USA) with a linear heating rate of 20 °C/min. Finally, the microstructure of powders was assessed by scanning electron microscopy (SEM, LSM5 Pascal; Zeiss, Oberkochen, Germany) according to González et al. [1].

3. Results and Discussion

Many authors have highlighted the influence of parent emulsion characteristics on the final properties of powders, especially particle size distribution [9], which is useful for assessing the homogeneity of the system and tracking changes caused by processing transformations. Tiny and agglomerated oil droplets could be identified in emulsions (Figure 1) after complex coacervation of SPI and GA, which enhances the microparticles' mechanical strength [10]. As regards the rheological behavior of parent emulsions, the data were fit to the power law ($R^2 > 0.96$). Flow index (n) values in the range of 0.32–0.40 were found, indicating a shear-thinning behavior. The consistency index (k) was in the range of 2.40–3.00 Pa s^n (0.1–300 s^{-1} shear rate range). Finally, the viscosity value at 100 s^{-1} (η_{100}), typical of many food processes, fell in the range of 0.111–0.155 Pa s.

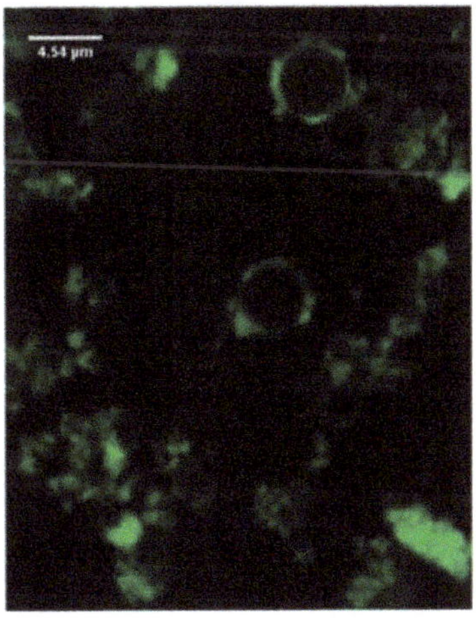

Figure 1. Confocal laser scanning microscopy (CLSM) images of parent chia seed-oil-in-water emulsions.

Ranges for the particle size distribution of parent and reconstituted emulsions, as well as for the corresponding AI of powders, are shown in Table 1. In addition, the corresponding ranges for other physical properties can be found in Table 2 (MC, a_w, SF, EE, CI and HR) and Table 3 (color parameters). The AI values for SF-SD microparticles were significantly higher ($p < 0.05$) than for TF-SD microparticles, which may be explained by the particle agglomeration facilitated by rotary atomizers [4]. In addition, a strong positive correlation ($p < 0.05$) was found between AI and d_{43} (r = 0.9916).

Table 1. Particle size parameters of parent and reconstituted emulsions.

	d_{43}	d_{32}	d_{50}	AI
Parent emulsion	21.10–21.20	12.80–12.90	12.21–12.30	--
T-F SD [A]	8.70–16.80	8.40–11.25	8.50–11.46	0.06–0.31
S-F SD [B]	45.30–60.00	9.50–13.20	48.50–51.70	3.00–4.30

[A] TF-SD (Tall-form spray dryer); [B] SF-SD (Short-form spray dryer). These abbreviations hold for the rest of the Tables. d_{43} (de Brouckere mean diameter, μm); d_{32} (Sauter mean diameter, μm); d_{50} (median diameter, volume distribution, μm); AI (aggregation index). The lowest and the highest values in each range correspond to the maximum air inlet temperature and the minimum feed flow rate, and to the minimum air inlet temperature and the maximum feed flow rate, respectively.

Table 2. Moisture content, water activity, surface fat, encapsulation efficiency and flowing properties of powders.

	MC	a_w	SF	EE	CI	HR
T-F SD	3.10–4.50	0.270–0.335	7.00–10.50	61.50–75.00	40.20–46.30	1.45–2.00
S-F SD	5.08–6.55	0.320–0.385	0.25–2.10	92.40–99.00	31.50–49.60	1.65–1.81

MC (moisture content, % wet basis); a_w (water activity); SF (surface fat, % dry basis); EE (encapsulation efficiency, % dry basis); CI (Carr's Index, %); HR (Hausner ratio).

Table 3. Color parameters of powders.

	L*	a*	b*	ΔE	WI	YI
T-F SD	91.10–94.05	−0.19–0.10	9.50–12.40	2.30–3.65	54.00–65.45	16.80–19.50
S-F SD	79.50–86.70	−0.21–1.10	14.20–18.05	7.40–13.50	32.40–44.05	23.40–29.80

L* (lightness); a* (red-green component); b* (yellow-blue component); ΔE (color change index); WI (whiteness index); YI (yellowness index).

A significant reduction ($p < 0.05$) in OSI values was observed for SF-SD powders, compared with bulk CSO: 0.21–0.80 h and 3.00–3.33 h, respectively. This may be related to eddies around the air disperser, which created local areas of counter-current flow between spray and air. On the other hand, the plug-flow air conditions in the chamber of a TF-SD are considerably different from the swirling air motion around the central disperser of the SF-SD, with less back mixing [4], thus preserving CSO from thermo-oxidative damage. For the TF-SD, the OSI values were in the range of 5.10–6.60 h.

The lowest and the highest values in each range correspond to the maximum air inlet temperature and the minimum feed flow rate, and to the minimum air inlet temperature and the maximum feed flow rate, respectively.

Strong correlations ($p < 0.05$) were also found between OSI and WI (r = 0.9358), L* (r = 0.9079), YI (r = −0.9029) and ΔE (r = −0.9410). Color changes are a sign of the oxidative deterioration of triacylglycerols and fatty acids, since this process can yield colored products [8]. In this regard, it is important to highlight that blank microparticles showed L*, a*, b*, WI and YI values in the ranges of 90.80–92.00, −0.03–0.05, 8.80–8.90, 65.40–65.50 and 13.70–13.80, respectively. Significant differences were observed ($p < 0.05$) between blank and SF-SD microparticles, as reflected by their higher values for ΔE compared to TF-SD powders, which was in consonance with greater oxidative damage.

The lowest and the highest values in each range correspond to the maximum air inlet temperature and the minimum feed flow rate, and to the minimum air inlet temperature and the maximum feed flow rate, respectively.

The particle size distributions and color parameters of the powders were also correlated ($p < 0.05$), as expected: WI–d_{43} (r = −0.9540) and WI–AI (r = −0.9627). Finally, a higher EE was associated with greater values of d_{43} (r = 0.9283) and AI (r = 0.9543).

As regards the flowing properties, CI > 25% and HR > 1.4 were observed, evidencing a poor flowability and strong inter-particle forces [8].

SEM micrographs of powders (Figure 2) showed tiny agglomerates, especially for a TF-SD, which are typically produced by twin-fluid atomizers and aid in the redispersion process [4]. On the other hand, round-shaped particles of greater size and no evident agglomerates were observed for the SF-SD.

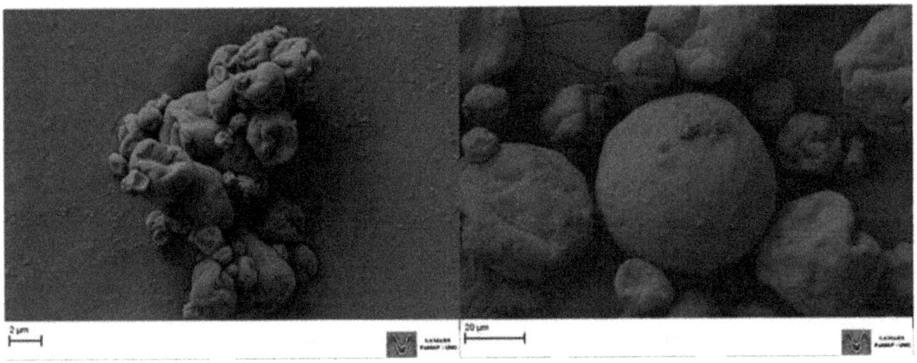

Figure 2. Scanning electron microscopy (SEM) images for microencapsulated chia seed oils. (**First image**) TF-SD. (**Second image**) SF-SD.

The thermal performance of the powders compared with a blank microparticle was assessed. It was found that all formulations showed two main stages of mass loss (curves not shown). The first stage, below 100 °C, was related to the loss of adsorbed and bound water, and had a small weight loss for all microparticles. The onset temperatures for the second degradation stage were around 250–260 °C for all samples. These temperatures were in consonance with values reported for the wall materials of the present work, and corresponded to SPI pyrolysis (~270 °C, Song et al. [11]) and the thermal degradation profile of polysaccharides (Castro-Cabado et al. [12]). Finally, glass transition temperatures (Figure 3), as shown by the DSC curves, were around 172, 169 and 186 °C for blank, TF-SD and SF-SD microparticles, respectively, in accordance with the reported values for MD DE 5 [3], GA (Barros Fernandes et al. [13]) and SPI (Tang et al. [14]).

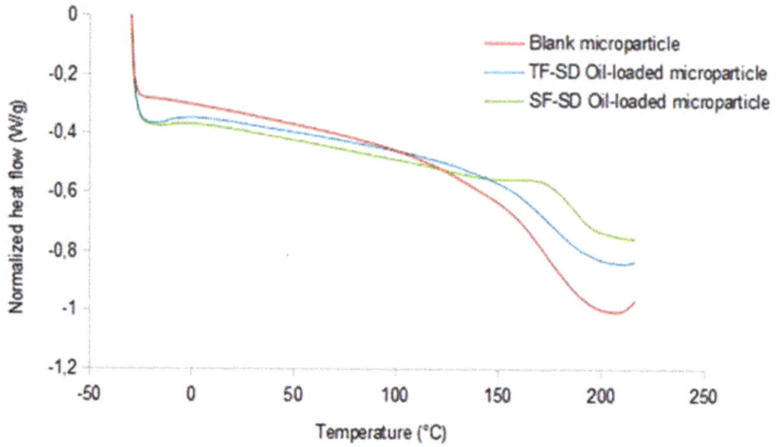

Figure 3. Differential scanning calorimetry (DSC) curves for blank and chia seed oil-loaded microparticles.

4. Conclusions

The physico-chemical properties of microencapsulated CSO proved to be greatly affected by the spray dryer's aspect ratio and by the spray-air contact. Turbulence around the air disperser in the SF-SD created local areas of counter-current flow between spray and air, damaging the oil chemical quality. L*, WI, YI and ΔE values varied accordingly. However, a greater EE was achieved in the same SF-SD, which was associated with higher d_{43} and AI values. It can be concluded that a co-current spray–air contact constitutes a better alternative for the protection of a heat-sensitive ingredient like CSO.

Author Contributions: (M.G.B & N.C.): methodology, investigation; (M.G.B, N.P.X.A, M.C.P., V.D.T., M.L.M. & P.D.R.): methodology, conceptualization, formal analysis; (M.G.B & M.L.M.): writing; (V.D.T., M.L.M. & P.D.R.): supervision, project administration, funding acquisition.

Acknowledgments: The authors would like to acknowledge the Iberoamerican Project CYTED 119RT0567.

Funding: This work was supported by grant Ia ValSe-Food-CYTED (Ref. 119RT0567), Fondo para la Investigación Científica y Tecnológica (FONCyT, BID PICT 2014-2283, and PICT 2016-1150) and SeCyT-UNC.

Conflicts of Interest: The authors declare no conflict of interest.

References

1. González, A.; Martínez, M.; León, A.E.; Ribotta, P.D. Effects on bread and oil quality after functionalization with microencapsulated chia oil. *J. Sci. Food Agric.* **2018**, *98*, 4903–4910, doi:10.1002/jsfa.9022.
2. Timilsena, Y.P.; Adhikari, R.; Barrow, C.J.; Adhikari, B. Microencapsulation of chia seed oil using chia seed protein isolate chia seed gum complex coacervates. *Int. J. Biol. Macromol.* **2016**, *91*, 347–357, doi:10.1016/j.ijbiomac.2016.05.058.
3. Anandharanakrishnan, C.; Ishwarya, S. *Spray Drying Techniques for Food Ingredient Encapsulation*; John Wiley & Sons, Ltd.: Chichester, UK, 2015; ISBN: 978-1-118-86419-7.
4. Masters, K. *Spray Drying Handbook*, 3rd ed.; Chapter 1; Halsted Press: New York, NY, USA; John Wiley & Sons Inc.: Hoboken, NJ, USA, 1979. Available online: https://trove.nla.gov.au/work/6866065 (accessed on 20 May 2019).
5. Martínez, M.; Marin, M.; Salgado, C.; Revol, J.; Penci, M.; Ribotta, P. Chia (*Salvia hispanica* L.) oil extraction: Study of processing parameters. *LWT–Food Sci. Technol.* **2012**, *47*, 78–82, doi:10.1016/j.lwt.2011.12.032.

6. González, A.; Gastelú; G; Barrera, G.N.; Álvarez Igarzabal, C.I. Preparation and characterization of soy protein films reinforced with celloluse nanofibers obtained from soybean by-products. *Food Hydrocoll.* **2019**, *89*, 758–764, doi:10.1016/j.foodhyd.2018.11.051.
7. Ma, T.; Zhao, H.; Wang, J.; Sun, B. Effect of processing condition on the morphology and oxidative stability of lipid microcapsules during complex coacervation. *Food Hydrocoll.* **2019**, *87*, 637–643, doi:10.1016/j.foodhyd.2018.08.053.
8. Rodríguez, E.S.; Julio, L.M.; Henning, C.; Diehl, B.W.K.; Tomás, M.C.; Ixtaina, V.Y. Effect of natural antioxidants on the physicochemical properties and stability of freeze-dried microencapsulated chia seed oil. *J. Sci. Food Agric.* **2018**, doi:10.1002/jsfa.935.
9. Us-Medina, U.; Julio, L.M.; Segura-Campos, M.R.; Ixtaina, V.Y.; Tomás, M.C. Development and characterization of spray-dried chia oil microcapsules using by-products from chia as wall material. *Powder Technol.* **2018**, *334*, 1–8, doi:10.1016/j.powtec.2018.04.060.
10. Ach, D.; Briançon, S.; Broze, G.; Puel, F.; Rivoire, A.; Galvan, J.-M.; Chevalier, Y. Formation of microcapsules by complex coacervation. *Can J Chem Eng.* **2015**, *93*, 183–192, doi:10.1002/cjce.22086.
11. Song, X.; Zhou, C.; Fu, F.; Zhilin, C.; Qinling, W. Effect of high-pressure homogenization on particle size and film properties of soy protein isolate. *Ind. Crops Prod.* **2013**, *43*, 538–544, doi:10.1016/j.indcrop.2012.08.005.
12. Castro-Cabado, M.; Parra-Ruiz, F.J.; Casado, A.L.; San Román, J. Thermal crosslinking of maltodextrin and citric acid. Methodology to control the polycondensation reaction under processing conditions. *Polym. Polym. Compos.* **2016**, *24*, 643–653, doi:10.1177/096739111602400803.
13. de Barros Fernandes, R.V.; Borges, S.V.; Botrel, D.A. Gum arabic/starch/maltodextrin/inulin as wall materials on the microencapsulation of rosemary essential oil. *Carbohydr. Polym.* **2013**, *101*, 524–532, doi:10.1016/j.carbpol.2013.09.083.
14. Tang, C.-H.; Chen, Z.; Li, L.; Yang, X.-Q. Effects of transglutaminase treatment on the thermal properties of soy protein isolates. *Food Res. Int.* **2006**, *39*, 704–711, doi:10.1016/j.foodres.2006.01.012.

proceedings

MDPI

Proceedings

Development and Characterization of Functional O/W Emulsions with Chia Seed (*Salvia hispanica* L.) by-Products †

Luciana M. Julio, Vanesa Y. Ixtaina and Mabel C. Tomás *

Centro de Investigación y Desarrollo en Criotecnología de Alimentos (CIDCA), CONICET, CICPBA, Universidad Nacional de La Plata (UNLP), 47 y 116, 1900 La Plata, Argentina; luci_julio86@hotmail.com (L.M.J.); vanesaix@hotmail.com (V.Y.I.)

* Correspondence: mabtom@hotmail.com or mtomas@quimica.unlp.edu.ar

† Presented at the 2nd International Conference of la ValSe-Food Network, Lisbon, Portugal, 21–22 October 2019.

Published: 1 September 2020

Abstract: Physicochemical properties of O/W emulsions containing functional ingredients (high ω-3 fatty acid content, protein, and soluble fiber) from chia seeds with different protein–carbohydrate combinations (sodium caseinate-lactose, sodium caseinate-maltodextrin, and chia protein-rich fraction-maltodextrin) and chia mucilage were studied. Sodium caseinate with lactose or maltodextrin produced O/W emulsions with small droplet size, high uniformity in droplet size distribution, negatively charged droplets (pH 6.5), pseudoplastic behavior, and high physical stability. Emulsions with chia protein-rich fraction presented wider droplet size distribution and higher $D_{3.2}$ values than the previous ones, recording a Newtonian behavior. The addition of chia mucilage affected the rheological characteristics of emulsions.

Keywords: chia by-products; chia mucilage; O/W emulsions; ω-3 fatty acids; chia protein-rich fractions

1. Introduction

The demand for functional foods with multiple health benefits has increased in recent years due to the new trend towards a healthy lifestyle. Functional foods are designed to supply basic nutrients as well as to reduce the risk of some diseases. Advances in food technology resulting in new components, products, processes, and packaging have provided more opportunities for value-added products.

Nowadays, interest in the substitution of synthetic emulsifiers and stabilizers by others of natural origin, such as vegetal polysaccharides and proteins, has grown. The industry generates residual cake with a high fiber and protein content after the extraction of chia oil from the seeds. These by-products are mainly used for animal feed with limited economic and social impacts. Thus, an alternative to adding value to these by-products would be the application of technologies to develop functional food that including them.

A chia protein-rich fraction containing 64.9% of globulins, 20.2% of glutelins, 10.9% of albumins, and 4.0% of prolamins was studied by Sandoval-Oliveros and Paredes-López [1]. This protein-rich fraction presented high contents of glutamic acid, arginine, and aspartic acid, which are important for the proper functioning of the immune system and the prevention of cardiovascular diseases. Besides, chia seed contains between 5 and 6% of mucilage, a tetrapolysaccharide of high molecular weight mainly composed of D-xylose, D-mannose, D-arabinose, D-glucose, and galacturonic and glucuronic acids [2,3]. The intake of chia mucilage as dietary fiber source was associated with numerous health benefits, including the reduction of the risk of coronary heart disease, diabetes, obesity, and different types of cancer [4].

This work deals with the development and characterization of oil-in-water (O/W) emulsions with chia oil, evaluating the influence of different combinations of proteins–carbohydrates and the application of chia by-products (mucilage, protein-rich fraction) on the physicochemical properties of these systems.

2. Materials and Methods

2.1. Material

Chia oil ($C_{16:0}$ 9.27%; $C_{18:0}$ 3.41%; $C_{18:1}$ 9.37%; $C_{18:2}$ 17.58%; $C_{18:3}$ 59.02%; $C_{20:0}$ 1.36%) was provided by SDA S.A. (Argentina). Casein sodium from bovine milk was purchased from Sigma Chemical Company (St. Louis, MO), the Maltodextrin DE 13–17% was obtained from Productos de Maíz S.A. (Argentina) and the D-lactose monohydrate from Anedra (Argentina). All reagents used were of analytical grade.

Chia protein-rich fraction with 43.0 of protein, 0.7 of fat, 8.4 of moisture, 14.1 of fiber, 8.4 of ash, and 25.4% of nitrogen-free extract was obtained by a dry processing of defatted chia flour according to Vázquez-Ovando et al. [5]. Chia mucilage was obtained from whole chia seeds according to Segura-Campos et al. [6] method with some modifications. The proximal composition of chia mucilage was 10.7, 8.9, 9.1, 3.9, 13.6, and 53.8% of moisture, ash, protein, fat, fiber, and nitrogen-free extract, respectively.

2.2. Methods

2.2.1. Preparation of Emulsions

Oil-in-water (O/W) emulsions were prepared mixing 10% (*wt/wt*) of chia oil and 90% (*wt/wt*) of aqueous phase with different compositions (Table 1) using a rotor–stator system Ultraturrax T-25 (Janke and Kunkel GmbH, Staufen, Germany) at 9500 rpm, 1 min. Then, in a second homogenization stage, the samples were passed four times through a high-pressure homogenizer (Panda 2 K, GEA NiroSoavi, Parma, Italy) at 600 bar. Nisine 0.0012% (*wt/wt*) and potassium sorbate 0.1% (*wt/wt*) were added to the emulsions to prevent microbial growth. Emulsions were stored at 4 ± 1 °C and protected from light for 15 days.

2.2.2. Droplet Size

The droplet size distribution and the De Sauter ($D_{3,2}$) mean diameter were obtained using a laser diffraction Malvern Mastersizer 2000E particle size analyzer (Malvern Mastersizer 2000E, Malvern Instruments Ltd., Worcestershire, UK) in a range of 0.1–1000 μm.

2.2.3. ζ-potential

The ζ-potential was determined using a Zeta Potential Analyzer (Brookhaven 90Plus/Bi-MAS, USA) instrument at room temperature according to Julio et al. [7]. The ζ-potential range was set from −100 to 50 mV. For each determination, 50 mg of emulsion was dispersed in 100 mL of milli-Q water.

Table 1. Composition of chia oil-in-water (O/W) emulsions.

Sample	Chia Oil	Aqueous Phase Composition % *wt/wt*				
		Chia Protein-Rich Fraction	**Sodium Caseinate**	**Lactose**	**Malodextrin**	**Chia Mucilage**
CL	10	-	10	10	-	-
CM	10	-	10	-	10	-
PM	10	10	-	-	10	-
CL + Mg	10	-	10	10	-	0.2
CM + Mg	10	-	10	-	10	0.2
PM + Mg	10	10	-	-	10	0.2

Chia protein-rich fraction (P), sodium caseinate (C), lactose (L), maltodextrin (M), and chia mucilage (Mg).

2.2.4. Rheological Properties

Rheological measurements were performed using a Haake RS600 controlled stress oscillatory rheometer (Haake, Germany) with a coarse plate–plate sensor system at 25 ± 1 °C. The samples were subjected to a logarithmic increasing of shear rate from 1 to 500 s^{-1} in 2 min, followed by a steady shear at 500 s^{-1} for 1 min, and finally a decreasing shear rate from 500 to 1 s^{-1} in 2 min [8].

2.2.5. Emulsion Stability

Physical stability of emulsions was determined by measurements of dispersed light using a Vertical Scan Analyzer Quick Scan (Coulter Corp., Miami, FL, USA) according to Pan et al. [9]. The emulsions were transferred to cylindrical glass tubes and periodically measured during 15 days.

3. Results and Discussion

The droplet size distribution (DSD) of chia O/W emulsions with sodium caseinate presented DSD curves with a mono (CL) or bimodal (CM) shape. On the other hand, emulsions prepared with chia protein-rich fraction exhibited wider and trimodal DSD with a shift towards larger droplet sizes. Additionally, systems with chia mucilage addition presented similar DSD curves shape but shifted to lower particle sizes, which was especially noticeable for emulsions with the protein-rich fraction.

The protein–carbohydrate combination had a significant effect ($p \leq 0.05$) on the $D_{3,2}$ diameter of emulsions droplets. At the initial time, emulsions with sodium caseinate presented droplet sizes between 0.22 and 0.27 μm and Span values from 1.07 to 1.46, exhibiting a high degree of uniformity. Emulsions containing the protein-rich fraction presented droplet sizes of ~9.86 μm and Span values from 2.23 to 2.65 due to the presence of bigger particle populations. The larger drop size of the emulsions of chia protein-rich fractions could be due to their lower level of protein available under the conditions of the chemical environment. Thus, when the emulsifying agent is not enough to fully stabilize the droplet interface, larger particles may be formed during homogenization [10,11]. Additionally, since sodium caseinate (protein with structural flexibility) is more effective to reduce the interfacial tension at the interface than chia proteins, which are mainly constituted by globular proteins [1], it is expected that it plays a major role as an emulsifier. Besides, a smaller ($p \leq 0.05$) droplet size for PM + Mg emulsions (7.44 μm) in comparison with PM systems (9.86 μm) was observed. This fact could be due to the increase of viscosity in systems with mucilage, which reduces the movement of the oil droplets, their collision, and coalescence.

The surface droplet charge at pH 6.5 was negative for all O/W emulsions, probably due to the ionized groups of the proteins at pH above the isoelectric point (pI) (Figure 1). The electric charge of the emulsion droplets stabilized with sodium caseinate was −35 and −31 mV for CL and CM, respectively, while those coated by the chia protein-rich fraction resulted in ~−23 mV. The more negative charge in droplets of protein-rich fraction systems could be due to the presence of anionic functional groups present in the chia protein structure, mainly related to glutamic and aspartic acids. Besides, the net charge of oil droplets became less ($p \leq 0.05$) negative when chia mucilage was added into the emulsions, possibly related to the charge suppression caused by electrostatic associations between polypeptide chains and charged groups of the chia mucilage.

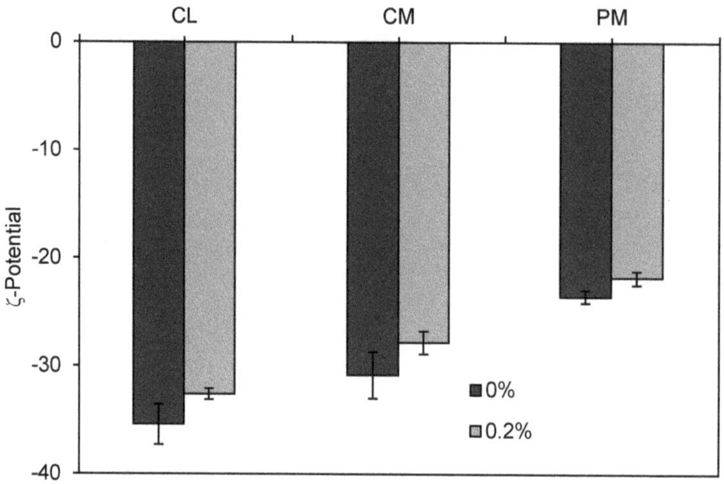

Figure 1. ζ-potential values of chia O/W emulsions at pH 6.5.

Regarding rheological properties, emulsions were affected by the proteins–carbohydrates combination used and the addition of chia mucilage. The experimental data, corresponding to rheological measurements, was fitted to the power-law model, and n (flow behavior index) and K (consistency coefficient) parameters were calculated. Differences in the flow behavior of the different O/W emulsions were evidenced (Figure 2). Systems with sodium caseinate recorded values of $n < 1$, exhibiting pseudo-plasticity on different levels. In this sense, emulsions with maltodextrin had greater pseudoplastic behavior than lactose ones. On the other hand, emulsions with the chia protein-rich fraction presented a Newtonian behavior ($n \sim 1$) (Figure 2).

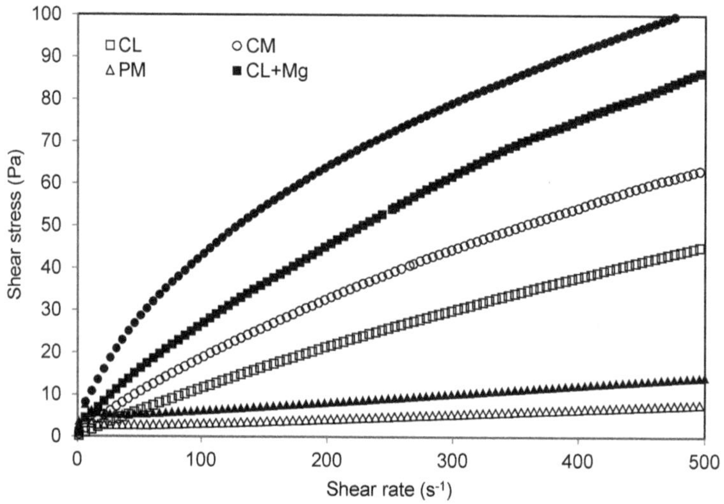

Figure 2. Flow curves of chia O/W emulsions. Average values (n = 3).

The apparent viscosity of emulsions at 100 s^{-1} (η_{100}), typical of food processes such as flow through pipes, agitation, and chewing [12], was also calculated. In this sense, systems with the chia protein-rich fraction had lower η_{100} values ($p \leq 0.05$) than emulsions stabilized with sodium caseinate. This fact could be related to a significant amount of nonadsorbed sodium caseinate in the continuous phase, which would lead to the formation of aggregates and a transient network structure with an enhancement in the viscosity [7]. Furthermore, there was an increase ($p \leq 0.05$) in the viscosity of emulsions with sodium caseinate containing chia mucilage. Similar results were reported by Timilsena et al. [3], who attributed the high viscosity of the chia mucilage solutions to the presence of 4-O-methyl-glucuronic acid and the intermolecular chain networks formation in aqueous media.

The physical stability of each emulsion was examined through its optical characterization during 15 days. The backscattering (BS) profiles evolution, as a function of the sample height (10–20 and 40–50 mm) and the storage time, for the different systems, are presented in Figure 3. Emulsions with sodium caseinate had high physical stability, causing their BS profiles to remain unchanged during the entire storage period (Figure 3a,b). This behavior could be due to the high viscosity level and small particle size of these systems, which reduce the droplet mobility and therefore its upward movement according to Stokes's law. In contrast, emulsions prepared with the chia protein-rich fraction recorded clarification at the bottom of the sample tube at day 5 of storage (Figure 3a,b). This creaming occurrence would be caused by higher mobility and interaction of the oil droplets as a result of weak viscous forces in the aqueous phase of these systems. Emulsions with the chia mucilage addition presented similar BS profiles to those without this by-product.

(a)

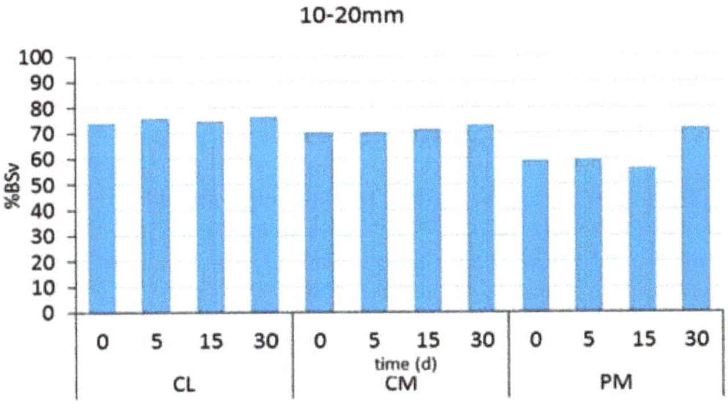

(b)

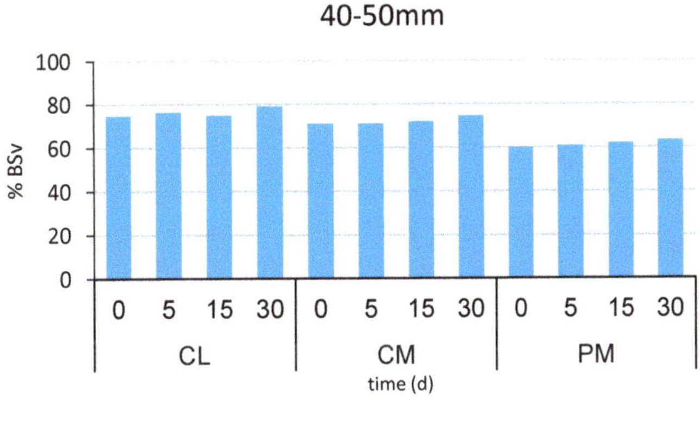

Figure 3. Backscattering profiles vs. sample tube height: (**a**) 10–20 mm and (**b**) 40–50 mm. Average values ($n = 2$). Std. deviation < 5%.

4. Conclusions

O/W emulsions proved to be suitable systems to deliver and protect chia seed by-products. All protein–carbohydrate combinations used for emulsions preparation led to the improvement of the oxidative stability of chia oil.

Chia mucilage addition had a significant effect on the rheological properties of the emulsions. Systems containing chia mucilage recorded higher viscosity and global stability due to the reduction of the oil droplets movement. Thus, chia mucilage exhibited a potential role as a thickening agent.

The obtained information could be applied to design and develop O/W emulsions as delivery systems of ω-3 fatty acids and other by-products from chia seed, allowing the revaluation of these novel ingredients.

Acknowledgments: This work was supported by grant Ia ValSe-Food-CYTED (119RT0567), Proyecto X756, X907 Universidad Nacional de La Plata and PICT 2016 0323 ANPCyT (Argentina).

References

1. Sandoval-Oliveros, M.R.; Paredes-López, O. Isolation and characterization of proteins from chia seeds (*Salvia hispanica* L.). *J. Agric. Food Chem.* **2012**, *61*, 193–201, doi:10.1021/jf3034978.
2. Reyes-Caudillo, E.; Tecante, A.; Valdivia-López, M. Dietary fibre content and antioxidant activity of phenolic compounds present in Mexican chia (*Salvia hispanica* L.) seeds. *Food Chem.* **2008**, *107*, 656–663, doi:10.1016/j.foodchem.2007.08.062.
3. Timilsena, Y.P.; Adhikari, R.; Kasapis, S.; Adhikari, B. Molecular and functional characteristics of purified gum from Australian chia seeds. *Carbohydr. Polym.* **2016**, *136*, 128–136, doi:10.1016/j.carbpol.2015.09.035.
4. Vázquez-Ovando, A.; Betancur-Ancona, D.; Chel-Guerrero, L. Physicochemical and functional properties of a protein-rich fraction produced by dry fractionation of chia seeds (*Salvia hispanica* L.). *CyTA-J. Food* **2012**, *11*, 75–80, doi:10.1080/19476337.2012.692123.
5. Mann, J. I.; Cummings, J. H. Possible implications for health of the different definitions of dietary fibre. *Nutr. Metab. Cardiovas.* ***2009***, 19, 226–229. doi:10.1016/j.numecd.2009.02.002
6. Segura-Campos, M.R.; Ciau-Solís, N.; Rosado-Rubio, G.; Chel-Guerrero, L.; Betancur-Ancona, D. Chemical and functional properties of chia seed (*Salvia hispanica* L.) gum. *Int. J. Food Sci.* **2014**, *2014*, doi:10.1155/2014/241053.
7. Julio, L.M.; Ixtaina, V.Y.; Fernández, M.A.; Sánchez Torres, R.M.; Wagner, J.R.; Nolasco, S.M.; Tomás, M.C. Chia seed oil-in-water emulsions as potential delivery systems of ω-3 fatty acids. *J. Food Eng.* **2015**, *162*, 48–55, doi:10.1016/j.jfoodeng.2015.04.005.
8. Capitani, M.; Spotorno, V.; Nolasco, S.M.; Tomás, M.C. Physicochemical and functional characterization of by-products from chia (*Salvia hispanica* L.) seeds of Argentina. *LWT-Food Sci. Technol.* **2012**, *45*, 94–102, doi:10.1016/j.lwt.2011.07.012.
9. Pan, L.; Tomás, M.; Añon, M. Effect of sunflower lecithins on the stability of water-in-oil and oil-in-water emulsions. *J. Surfactants Deterg.* **2002**, *5*, 135–143, doi:10.1007/s11743-002-0213-1.
10. Day, L.; Xu, M.; Hoobin, P.; Burgar, I.; Augustin, M. Characterisation of fish oil emulsions stabilised by sodium caseinate. *Food Chem.* **2007**, *105*, 469–479, doi:10.1016/j.foodchem.2007.04.013.
11. Dickinson, E. Hydrocolloids at interfaces and the influence on the properties of dispersed systems. *Food Hydrocoll.* **2003**, *17*, 25–39, doi:10.1016/S0268-005X(01)00120-5.
12. McClements, D.J. *Food Emulsions: Principles, Practices, and Techniques*; CRC Press: Boca Raton, FL, USA, 2004; ISBN 0849320232.

proceedings

Proceedings

A Comparative Study of the Physical Changes of Two Soluble Fibers during In Vitro Digestion †

Natalia Vera [1,2], Laura Laguna [3], Liliana Zura [2] and Loreto A. Muñoz [2,*]

1 Departamento de Ciencias de los Alimentos y Tecnología Química, Universidad de Chile, Santos Dumont 964, Independencia, Santiago 8380494, Chile; nati.vcespedes@gmail.com
2 Escuela de Ingeniería, Universidad Central de Chile, 8330601 Santiago, Chile; liliana.zura@gmail.com
3 Institute of Agrochemistry and Food Technology (IATA), C/Catedrático Agustín Escardino Benlloch, 7, 46980 Paterna, Spain; laura.laguna@iata.csic.es
* Correspondence: loreto.munoz@ucentral.cl

† Presented at the 2nd International Conference of Ia ValSe-Food Network, Lisbon, Portugal, 21–22 October 2019.

Published: 1 September 2020

Abstract: This research aimed to compare the apparent viscosity and the degree of fragmentation/aggregation produced in dispersions of xanthan gum and chia mucilage during the gastrointestinal tract by using an in vitro digestion. Both soluble fibers exhibited pseudoplastic behavior, independent of the concentration and stage of digestion (oral, gastric or intestinal). The viscosity decreased from the oral to intestinal stage in all the concentrations, produced mainly by the "dilution effect" by the addition of digestive fluids. The particle size of xanthan gum increased drastically in the gastric stage mainly due to the decrease in pH, but at intestinal level returned to its original pattern, while particle size and pattern of mucilage during all the stages of digestion remained unchanged, maintaining its integrity. In general terms, since chia mucilage and xanthan gum maintain their viscosity and integrity through the gastrointestinal tract, they could be used as functional ingredients improving the functionality of foods.

Keywords: chia seed; in vitro digestion; mucilage; soluble fiber; xanthan gum

1. Introduction

According to several authors and health organizations, diets high in dietary fiber have a higher incidence in the prevention of many major non-communicable diseases compared to diets lower in this component [1–5]. Many of the beneficial effects have been explained by their behavior at gastrointestinal level. According to EFSA (European Food Safety Authority), dietary fiber corresponds to non-digestible carbohydrates plus lignin, including all carbohydrate components occurring in foods that are non-digestible in the human small intestine and pass into the large intestine. Based on their solubility, dietary fiber can be classified into water soluble (pectins, gums, mucilages, etc.) and insoluble fractions (cellulose, lignin, etc.); both types have different molecular characteristics and physiological effects on the gastrointestinal tract [6]. Soluble dietary fiber (DSF) intake is generally associated with slow transit through the stomach and increasing of the small intestine transit time; this behavior is related with its ability to form viscous solutions [7]. In addition, soluble fibers are fermented by the colonic microbiota, releasing different levels of short chain fatty acids (SCFAs) and play a critical role in the composition and metabolic activity of the microbiome, which affects the intestinal health and ultimately the immune system and the body's ability to resist some chronic diseases [8].

Chia seed has been described as a source of soluble fiber, its mucilage has the ability to retain large amounts of water and produce viscous dispersions, even at low concentrations [9]. On the other hand, xanthan gum is a commercial soluble fiber obtained by fermentation of *Xanthomonas campestris*, is soluble in cold water and in solution exhibits high pseudoplastic flow [10].

The objective of this study was to evaluate comparatively the viscosity and degree of aggregation/fragmentation changes produced during the in vitro digestion of mucilage from chia seeds and xanthan gum.

2. Materials and Methods

2.1. Materials

Chia seeds were provided by Benexia (Functional Products Trending S.A., Santiago, Chile); the crude mucilage was obtained from chia seeds by using the method proposed by Muñoz et al. (2012) and xanthan gum was purchased from Sigma-Aldrich [11]. To perform the in vitro digestion, the enzymes and reagents were purchased from Sigma-Aldrich and Merck. The comparisons among means were performed using one-way ANOVA (Analysis of Variance) and the significant differences were determined by the Tukey test ($p < 0.05$). All these tests were performed using the software, Statgraphics Centurion XV.I.

2.2. In Vitro Digestion

Suspensions of the two DSF at low, medium and high concentrations (0.3, 0.5 and 1.0% *w/w*) were subjected to in vitro digestion simulating the gastrointestinal conditions (oral, gastric and intestinal). To perform the experiments, the standardized static in vitro digestion protocol proposed by Minekus et al. (2014) was used [12]. The simulated gastrointestinal fluids such as salivary (SSF), gastric (SGF) and intestinal (SIF) and the enzymes were prepared according to the same protocol.

2.3. Steady Shear Flow Behavior

The apparent viscosity to each soluble fiber dispersion was determined without digestion as control and before and after each digestion stage by applying an increasing shear rate from 0.1 to 100 s^{-1} in triplicate using a Rotational Rheometer, RheolabQC (Anton Paar GmbH, Austria-Europe). The rheometer was equipped with a double gap concentric cylinder and a Peltier temperature plate set at 37 °C, simulating body temperature. The flow behavior index (n) and consistency index (k) values were obtained by fitting to the Power Law model (Equation (1)):

$$\eta = k\gamma^{(n-1)} \quad (1)$$

where η is the shear viscosity (Pa s), k is the consistency index (Pa s^{-1}), γ is the shear rate (s^{-1}) and n is the fluid behavior index (dimensionless).

2.4. Determination of Degree of Aggregation/Fragmentation

The degree of aggregation/fragmentation was determined at each stage of in vitro digestion in terms of particle size distribution in six-fold, by laser light scattering with a Malvern Mastersizer 2000 (Malvern Instruments, Worcestershire, UK) software version 5.6, using water at 25 °C as solvent.

3. Results

3.1. Steady Shear Flow Behaviour

The steady flow behavior, consistency (k), and flow index (n) behavior of the mucilage and xanthan gum at low, medium, and high concentration, before and during each stage of in vitro digestion can be seen in Figure 1 and Table 1, respectively. All the dispersions show non-Newtonian behavior, with decreasing viscosity with increasing shear rate, also known as pseudoplasticity or shear-thinning behavior (Figure 1). The apparent viscosity to the samples without digestion (Figure 1a–c) show a directly proportional relationship with the concentration and k increase as concentration increased. Similar behavior was reported by Timilsena et al., (2015) where the rheological properties of the purified chia seed polysaccharide were evaluated [13]. Moreover, the apparent viscosity of mucilage and xanthan gum decreased from the oral to intestinal stage in all the concentrations (Figure 1d–l), caused mainly by the addition of digestive fluids (SSF, SGF and SIF) and was less affected by the pH changes and ionic strength. Similar behavior was previously observed by Lazaro et al. (2018) and Fabek et al. (2014) when different soluble fibers were subjected to in vitro digestion [14,15]. According to Vuksan et al. (2011), many of the beneficial physiological effects produced by the dietary fiber intake are associated with their capacity to hydrate and increase the viscosity of the human digesta. In this study, both DSF provided viscosity at gastrointestinal level, therefore, their use in food matrices would help to improve functionality [16].

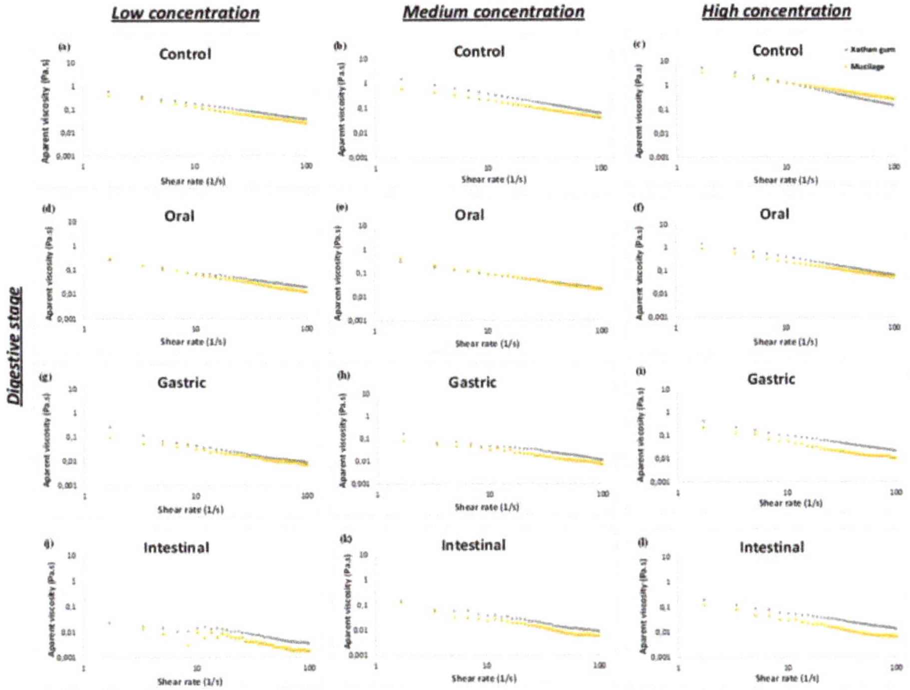

Figure 1. Apparent viscosity during in vitro digestion.

3.2. Degree of Aggregation/Fragmentation

Figure 2 shows the distribution of the particle size of the mucilage and xanthan gum during the in vitro digestion. Figure 2a–c shows the changes in particle size of xanthan gum at the three concentrations. In this case, xanthan gum did not show differences in the bimodal distribution between control and oral stage for the three concentrations, but at gastric level the particle size distribution changed; the curve moves to the right which indicates increasing size.

Table 1. Power of Law parameters for xanthan gum and mucilage from chia seed during in vitro digestion.

SDF	Concentration	Control (Before Digestion)			After Digestion								
					Oral			Gastric			Intestinal		
		n	*k*	R^2	*n*	*k*	R^2	*n*	*k*	R^2	*n*	*k*	R^2
Xanthan gum	Low	0.319 b	0.844 b	0.9994	0.399 a	0.318 b	0.9976	0.249 bc	0.254 ab	0.9880	0.333 a	0.074 b	0.9063
	Medium	0.219 bc	2.283 b	0.9998	0.376 a	0.391 b	0.9983	0.379 a	0.226 b	0.9710	0.351 b	0.185 a	0.9888
	High	0.081 d	9.479 b	0.9076	0.208 c	2.441 b	0.9506	0.314 a	0.507 b	0.9980	0.355 b	0.270 a	0.9960
Mucilage	Low	0.306 b	0.592 bc	0.9978	0.224 b	0.308 b	0.9970	0.201 c	0.320 a	0.9903	0.178 a	0.069 b	0.9406
	Medium	0.309 b	0.960 b	0.9989	0.363 a	0.235 b	0.9963	0.320 a	0.096 b	0.9839	0.337 b	0.130 a	0.9733
	High	0.339 b	5.173 b	0.9302	0.429 a	0.230 c	0.9991	0.379 a	0.180 c	0.9801	0.278 b	0.160 a	0.9836

Different letters signify difference across hydrocolloids for one concentration (low, medium or high); (*n*): flow behavior index and (*k*): consistency index values.

This behavior has been previously explained as the decrease of the intermolecular electrostatic repulsion at low pH [10], which allows the expansion of the fiber chains. Finally, at intestinal level, the particle size of xanthan gum returned to the original size mainly due to the increase in pH, at the same time as the concentration increases the conformation becomes monomodal. On the other hand, the particle size and pattern at the three concentrations of mucilage during all the stages of digestion remained unchanged (Figure 2d–f). This behavior has not been previously reported and could indicate a better physiological response when the mucilage from chia seed is ingested. Furthermore, the gastric empty will be slower by using mucilage than xanthan gum and possibly could reduce the nutrient absorption through the intestinal mucosa [17].

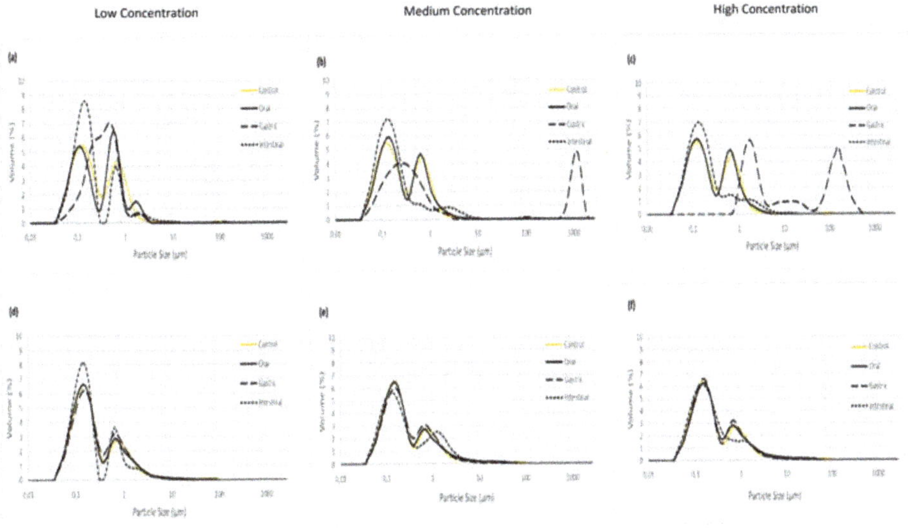

Figure 2. Degree of aggregation/fragmentation during in vitro digestion, (**a**–**c**) correspond to xanthan gum and (**d**–**f**) correspond to mucilage from chia seed.

4. Conclusions

In this study both soluble fibers, mucilage and xanthan gum, had the ability to retain viscosity through the gastrointestinal tract, which could indicate their ability to modulate certain physiological responses enhancing functionality when they are added into the food matrix. In addition, the aptitude of these fibers to maintain their structure suggests that they can be used to develop structured foods as a strategy to modulate the digestive process, delaying the gastric emptying.

Acknowledgments: This work was supported by the following grant: Ia ValSe-Food-CYTED (119RT0567); the National Fund for Scientific Development and Technological, Project FONDECYT 11150307, Chile and the Fundación para la Innovación Agraria PYT-2018-0261 FIA.

References

1. FAO (Food and Agriculture Organization); WHO (World Health Organization). *Report of a Joint FAO/WHO Expert Consultation, Diet, Nutrition and the Prevention of Chronic Disease*; Technical Report Series; FAO; WHO: Geneva, Switzerland 2003; No. 916 (TRS 916).
2. Chawla, R.; Patil, G. Soluble Dietary Fiber. *Compr. Rev. Food Sci. Food Saf.* **2010**, *9*, 178–196, doi:10.1111/j.1541-4337.2009.00099.x.
3. EFSA. Scientific Opinion on principles for deriving and applying Dietary Reference Values. *EFSA J.* **2010**, *8*, doi:10.2903/j.efsa.2010.1458.

4. Gidley, M.J. Hydrocolloids in the digestive tract and related health implications. *Curr. Opin. Colloid Interface Sci.* **2013**, *18*, 371–378, doi:10.1016/j.cocis.2013.04.003.
5. Capuano, E. The behavior of dietary fiber in the gastrointestinal tract determines its physiological effect. *Crit. Rev. Food Sci. Nutr.* **2016**, *57*, 3543–3564, doi:10.1080/10408398.2016.1180501.
6. Repin, N.; Cui, S.W.; Goff, H.D. Rheological behavior of dietary fibre in simulated small intestinal conditions. *Food Hydrocoll.* **2018**, *76*, 216–225, doi:10.1016/j.foodhyd.2016.10.033.
7. Taghipoor, M.; Barles, G.; Georgelin, C.; Licois, J.; Lescoat, P. Digestion modeling in the small intestine: Impact of dietary fiber. *Math. Biosci.* **2014**, *258*, 101–112, doi:10.1016/j.mbs.2014.09.011.
8. Li, Y.O.; Komarek, A.R. Dietary fibre basics: Health, nutrition, analysis, and applications. *Food Qual. Saf.* **2017**, *1*, 47–59, doi:10.1093/fqsafe/fyx007.
9. Tamargo, A.; Cueva, C.; Laguna, L.; Moreno-Arribas, M.V.; Muñoz, L.A. Understanding the impact of chia seed mucilage on human gut microbiota by using the dynamic gastrointestinal model simgi®. *J. Funct. Foods* **2018**, *50*, 104–111, doi:10.1016/j.jff.2018.09.028.
10. Brunchi, C.-E.; Bercea, M.; Morariu, S.; Dascalu, M. Some properties of xanthan gum in aqueous solutions: Effect of temperature and pH. *J. Polym. Res.* **2016**, *23*, 123, doi:10.1007/s10965-016-1015-4.
11. Munoz, L.; Cobos, A.; Díaz, O.; Aguilera, J. Chia seeds: Microstructure, mucilage extraction and hydration. *J. Food Eng.* **2012**, *108*, 216–224, doi:10.1016/j.jfoodeng.2011.06.037.
12. Minekus, M.; Alminger, M.; Alvito, P.; Balance, S.; Bohn, T.; Bourlieu, C.; Carrière, F.; Boutrou, R.; Corredig, M.; Dupont, D.; et al. A standardised staticin vitrodigestion method suitable for food—An international consensus. *Food Funct.* **2014**, *5*, 1113–1124, doi:10.1039/c3fo60702j.
13. Timilsena, Y.; Adhikari, R.; Kasapis, S.; Adhikari, B. Rheological and microstructural properties of the chia seed polysaccharide. *Int. J. Boil. Macromol.* **2015**, *81*, 991–999, doi:10.1016/j.ijbiomac.2015.09.040.
14. Lazaro, H.; Puente, L.; Zúñiga, M.C.; Muñoz, L.A. Assessment of rheological and microstructural changes of soluble fiber from chia seeds during an in vitro micro-digestion. *LWT* **2018**, *95*, 58–64, doi:10.1016/j.lwt.2018.04.052.
15. Fabek, H.; Messerschmidt, S.; Brulport, V.; Goff, H.D. The effect of in vitro digestive processes on the viscosity of dietary fibres and their influence on glucose diffusion. *Food Hydrocoll.* **2014**, *35*, 718–726, doi:10.1016/j.foodhyd.2013.08.007.
16. Vuksan, V.; Jenkins, A.L.; Rogovik, A.L.; Fairgrieve, C.D.; Jovanovski, E.; Leiter, L.A. Viscosity rather than quantity of dietary fibre predicts cholesterol-lowering effect in healthy individuals. *Br. J. Nutr.* **2011**, *106*, 1349–1352, doi:10.1017/s0007114511001711.
17. Bornhorst, G.M.; Kostlan, K.; Singh, R. Particle Size Distribution of Brown and White Rice during Gastric Digestion Measured by Image Analysis. *J. Food Sci.* **2013**, *78*, E1383–E1391, doi:10.1111/1750-3841.12228.

proceedings

MDPI

Proceedings

Microencapsulation of Chia Seed Oil (*Salvia hispanica* L.) in Spray and Freeze-Dried Whey Protein Concentrate/Soy Protein Isolate/Gum Arabic (WPC/SPI/GA) Matrices †

María Gabriela Bordón [1,2], Gabriela Noel Barrera [2,3], Maria C. Penci [1,2,3], Andrea Bori [4], Victoria Caballero [4], Pablo Ribotta [1,2,3] and Marcela Lilian Martínez [1,3,5,*]

1 Instituto de Ciencia y Tecnología de los Alimentos, Facultad de Ciencias Exactas, Físicas y Naturales (ICTA-FCEFyN)—Universidad Nacional de Córdoba (UNC), Córdoba 5000, Argentina; gabrielabordon90@gmail.com (M.G.B.); cpenci@gmail.com (M.C.P.); pribotta@agro.unc.edu.ar (P.R.)
2 Instituto de Ciencia y Tecnología de Alimentos Córdoba (ICYTAC, CONICET-UNC), Córdoba 5000, Argentina; gbarrera@agro.unc.edu.ar
3 Departamento de Química Industrial y Aplicada, FCEFyN-UNC, Córdoba 5000, Argentina
4 Estudiantes de Ingeniería Química, Departamento de Química Industrial y Aplicada, FCEFyN-UNC, Córdoba 5000, Argentina; andrea.bori.13@gmail.com (A.B.); victoriact68@hotmail.com (V.C.)
5 Instituto Multidisciplinario de Biología Vegetal (IMBIV, CONICET-UNC), Córdoba 5000, Argentina
* Correspondence: marcela.martinez@unc.edu.ar

† Presented at the 2nd International Conference of Ia ValSe-Food Network, Lisbon, Portugal, 21–22 October 2019.

Published: 1 September 2020

Abstract: Microencapsulation by different drying methods protects chia seed oil (CSO) against oxidative degradation, and ultimately facilitates its incorporation in certain foods. The aim of this work was to analyze the influence of freeze or spray drying, as well as of the coacervation phenomena in a ternary wall material blend—whey protein concentrate/soy protein isolate/gum arabic (WPC/SPI/GA)—on the physico–chemical properties of microencapsulated CSO. Differential scanning calorimetry studies indicated that the onset, peak, and end set temperatures for denaturation events shifted from 72.59, 77.96, and 78.02 to 81.34, 86.01, and 92.58 °C, respectively, in the ternary blend after coacervation. Oxidative stability indexes (OSI) of powders were significantly higher ($p < 0.05$) for both drying methods after inducing coacervation—from 6.45 to 12.04 h (freeze-drying) and 12.05 to 15.31 h (spray drying)—which was possibly due to the shifted denaturation temperatures after biopolymer interaction. It can be concluded that the ternary WPC/SPI/GA blend constitutes an adequate matrix to encapsulate CSO.

Keywords: chia seed oil; complex coacervation; freeze and spray drying; microencapsulation; oxidative stability; ternary wall material blend

1. Introduction

Chia seeds constitute a vegetable source rich in essential fatty acids given that, out of 30–40% total lipid (chia seed oil, CSO), 85% is composed of polyunsaturated fatty acids including Omega-3 (~66%) and Omega-6 (~20%). This composition entails a technological disadvantage due to its high susceptibility to oxidative degradation [1]. Therefore, suitable protection and delivery systems must be designed in order to preserve the functionality and organoleptic attributes of CSO within different formulated products [2].

Among microencapsulation methods, complex coacervation constitutes a great alternative for lipid cores due to the high encapsulation efficiency achieved [3]. It is an associative phase separation phenomenon where a protein and a polysaccharide exert strong attractive interactions between each other, mostly of an electrostatic nature [2]. Complex coacervates can be converted into powder either by freeze or spray drying, the latter being the most widespread technology in the microencapsulation field due to its low cost, flexibility and scalability [2,3]. Rapid water removal results in increased product shelf-life and provides the consumer with a stable product [4].

Soy protein isolate (SPI) has become one of the most utilized plant origin-emulsifiers, probably owing to the extensive production of this legume worldwide [5]. Nonetheless, it has been established that the emulsions formulated with these proteins alone are not very stable [2]. Therefore, coacervation between SPI and anionic polysaccharides has been studied to enhance the stability of emulsions [5]. In the present work, gum arabic (GA) was selected as the polysaccharide source for the emulsion formulation, mostly because of its dual role as both surfactant and drying matrix [4]. In addition, whey protein concentrate (WPC), besides its well-known emulsion capacity, contributes with an inherent antioxidant activity, owing to the formation of thick films at droplet interfaces and the chelation of prooxidant metals [4].

Based on the above exposed exposé, the present work aimed to analyze the influence of two different drying methods (freeze or spray drying) and the presence (or absence) of complex coacervation phenomena in a ternary wall material blend—whey protein concentrate/soy protein isolate/gum arabic (WPC/SPI/GA)—as an encapsulant matrix for CSO.

2. Materials and Methods

2.1. Materials

Chia seed oil (CSO) was extracted from seeds coming from the Salta province (Nutracéutica Sturla SRL, Argentina), as described by Martínez et al. [6], in a pilot plant screw press (Komet Model CA 59 G, IBG Monforts, Mönchengladbach, Germany). Soy protein isolate (SPI) SUPRO E with 90% protein on fat free basis was purchased from The Solae Company (Argentina); gum Arabic (GA) (Alland & Robert, Port-Mort, France) and acid whey protein concentrate (WPC) with 80% protein were purchased from local distributors (Distribuidora NICCO and Natural Whey, respectively; both from Argentina).

2.2. Preparation of WPC/SPI/GA Dispersions

Wall material powders were re-dispersed in Milli-Q water for 1 h with gentle stirring according to González et al. [1] and kept overnight at 4 °C for complete hydration. The necessary quantity of powder was weighted in order to yield a 30% (*w/v*) total solid concentration in dispersions, with a WPC/SPI/GA ratio (weight basis) of 8/1/1.

2.3. Zeta Potential and Turbidity Measurements

The charge density of the individual biopolymers' dispersions and their mixture (0.1% *w/v* total concentration) was determined in order to find the optimum pH (pH_{opt}), where the coacervate yield reaches its highest value and the net charge on the system is zero [2]. The measurements were performed at 25 °C (Nano Zetasizer, Worcestershire, Malvern Instruments, UK).

In addition, the turbidity of dispersions was measured at 600 nm by spectrophotometry in order to follow the formation of complex coacervate particles as a function of pH, according to Timilsena et al. [2].

2.4. Differential Scanning Calorimetry of Dispersions

Thermal analyses of individual biopolymers' dispersions and their mixture (with or without complex coacervation induced) were performed with a DSC823e Calorimeter (Columbus, OH, USA) according to Blanco-Canalis et al. [7]. Specific enthalpies (J/g dispersion), as well as onset, peak, and end set temperatures for protein denaturation events were informed.

2.5. Emulsion Preparation and Characterization

Coarse emulsions were prepared by the high speed homogenization of CSO and a mixture of WPC/SPI/GA (15,000 rpm, 2 min, Ultraturrax homogenizer IKA T18, (Janke & Kunkel GmbH, Staufen, Germany); WPC, SPI and GA were present in an 8/1/1 ratio (weight basis) in emulsions, which where formulated with wall material and CSO contents of 30 and 15% (*w*/*v*), respectively. The coarse emulsions were further homogenized in a high-pressure valve homogenizer at 700 bar (1 cycle, Emulsi Flex C5, Avestin, Ottawa, ON, Canada). Subsequently, fine emulsions were either dried (by spray or freeze-drying) or subjected to pH adjustment (pH = 4.0) in order to induce complex coacervation among biopolymers before drying. That pH value was determined as described in Section 2.3. The coacervation reaction was completed with stirring at room temperature. Particle size distribution of emulsions was determined according to Us-Medina et al. [8] and with a LA 950V2 Horiba (Kyoto, Japan) analyzer. Time-dependent steady shear properties of emulsions were evaluated using a controlled-stress rheometer MCR 301 Anton Paar, equipped with a plate-cone geometry (50 mm diameter) and working with a 0.05 mm gap [9]. The rheological data were fitted to the power law and the hysteresis area was calculated.

2.6. Spray and Freeze-Drying Processes

Spray drying of chia-oil-in-water emulsions was performed as described by González et al. [1] in a laboratory-scale spray dryer, Büchi B-290 (Büchi Labortechnik AG, Flawil, Switzerland) equipped with a two-fluid nozzle atomizer. The drying conditions were the following: 1050 L/h; drying air inlet temperature: 130 ± 1 °C; atomization air flow rate: 538 L/h; pump setting (feed volumetric flow rate)—20% (approximately 5.6 mL/min); aspirator setting (drying air volumetric flow rate)—100% (38 m^3/h). The emulsions were stored at −70 °C for 48 h before freeze-drying. This operation was performed in a laboratory bench-top freeze-dryer (L-T8, Rificor, Buenos Aires, Argentina) operated with a condenser at −50 °C and a vacuum of 0.1 mbar for 2 cycles of 24 h.

2.7. Powder Analysis

The powders obtained with both drying methods were characterized in terms of moisture content (MC), water activity (a_w), and particle size distribution (d_{43} volume-based mean diameter) as described by Us-Medina et al. [8]. The aggregation index (AI) of microparticles after drying of chia-oil-in-water emulsions was determined according to Ma et al. [10]. Color measurements were performed with a CM600d spectrophotometer (Konica Minolta, Tokyo, Japan) according to González et al. [1]. The whiteness (WI) and yellowness (YI) indexes were estimated as described by Rodriguez et al. [11]. Surface fat (SF) and encapsulation efficiency (EE) were determined according to González et al. [1]. The powder flowability was evaluated using the Carr's Index (CI) and the Hausner Ratio (HR) as described by Rodriguez et al. [11]. Finally, the morphology of complex coacervates and reconstituted emulsions was assessed with an optical microscope (Leica DM5000 B, Leica Microsystems, Wetzlar, Germany) operated in a dark field mode. The morphology of powders was evaluated by scanning electron microscopy (SEM, LSM5 Pascal; Zeiss, Oberkochen, Germany) according to González et al. [1].

3. Results and Discussion

The determination of zeta potential as a function of pH is a commonly used approach to study complex coacervation, given that the pH range at which the phenomenon occurs can be easily identified [1]. For the WPC/SPI/GA ternary blend, a rapid phase separation occurred (0.1%*w*/*v* total solids) at pH = 4.0, for which the charge density determined by zeta potential measurements was in the range of 0.659–1.100 mV, reaching practically zero net charge for the system. This was in consonance with turbidity determinations, for which a peak in absorbance values at 600 nm (1.360–

1.362) was observed at pH = 4.0, indicating the formation of insoluble complexes upon pH adjustment in a clear initial system (neutral pH, recorded absorbance values of 0.256–0.304). These complexes grow in size and can be readily identified as round-shaped entities by optical microscopy (Figure 1). Finally, an additional evidence of the interaction among biopolymers in the ternary blend was given by DSC studies (Table 1). The informed temperatures were in accordance with values reported elsewhere for whey proteins [12]. As can be seen, the onset, peak and end set temperatures for WPC denaturation shifted to higher ($p < 0.05$) values after complex coacervation, compared with a WPC dispersion and a WPC/SPI/GA mixture without induced coacervation. This may be related to the preferred biopolymer–biopolymer over biopolymer–water interactions during complex coacervation, preventing the complete hydration of proteins, and thus preserving them from denaturation [12].

Table 1. DSC parameters for wall material dispersions.

	ΔH J/g	T Onset °C	T Peak °C	T Endset °C
HAP-treated WPC [A]	1.10–1.30	71.30–72.40	76.50–77.70	81.30–81.50
WPC/SPI/GA mixture	0.74–0.96	71.60–72.60	77.40–78.00	84.00–84.30
WPC/SPI/GA mixture-c.c [B]	0.53–0.67	81.30–81.50	85.90–86.00	91.30–92.60

[A] HAP-treated WPC (WPC suspension homogenized at 700 bar in a high pressure valve homogenizer);
[B] cc (complex coacervation); ΔH (DSC enthalpy, J/g dispersion); T onset, T peak, T endset (°C).

Many authors pointed out the influence of parent emulsion characteristics on the final properties of powders [8], which are useful for tracking changes caused due to processing transformations. As regards particle size distribution, the emulsions showed d_{43} initial values of 5.00–5.40 μm and 12.10–12.80 μm after complex coacervation, which agrees with the droplet flocculation phenomena that take place during the multinucleated capsule formation. The rheological behavior data were fitted to the power law ($R^2 > 0.94$). Flow index (n) values of 0.61–0.67 were found for parent emulsions, indicating a shear-thinning behavior. The consistency index (k) values were 0.25–0.58 Pa s^n (0.1–300 s^{-1} shear rate range). Finally, the viscosity values at 100 s^{-1} (η_{100}), typical of many food processes, were 0.0613–0.1190 Pa s.

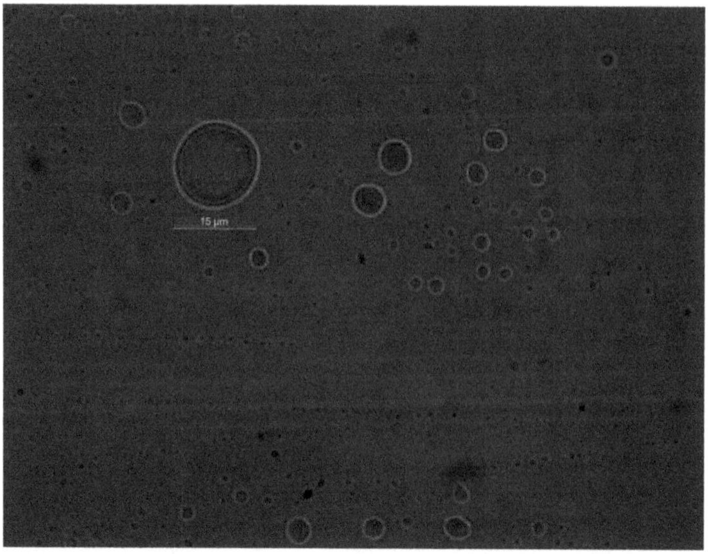

Figure 1. Optical microscopy images of whey protein concentrate/soy protein isolate/gum arabic (WPC/SPI/GA) complex coacervates.

The physico–chemical properties of powders can be found in Table 2. Values of d_{43} for reconstituted emulsions varied significantly ($p < 0.05$) according to the drying method: 51.60–61.20 μm (freeze drying) and 2.40–22.10 μm (spray drying). A significant correlation ($p < 0.05$, r = 0.7810) was found between d_{43} and AI, for which the highest values were observed in powders obtained by freeze drying.

This is in agreement with the fission forces that take place during atomization processes, reducing the particle size for enhanced heat and mass transfer [4] and yielding tiny solid particles (Figure 2). As expected, strong correlations ($p < 0.05$) were found for AI-WI (r = −0.9656) and AI-YI (r = 0.9694).

Table 2. Powder physico–chemical properties.

	MC	a_w	AI	EE	WI	YI	OSI
	% Wet Basis			**% Dry Basis**			**h**
FD [A]	3.70–4.30	0.216–0.220	17.00–39.70	65.60–77.70	18.70–31.40	30.30–36.70	6.45–12.04
SD [B]	2.91–4.01	0.214–0.257	0.14–0.16	71.30–80.07	61.30–64.10	15.20–16.22	12.05–15.31

[A] FD (freeze drying); [B] SD (spray drying); MC (moisture content, % wet basis); a_w (water activity); AI (aggregation index); EE (encapsulation efficiency, % dry basis); WI (whiteness index); YI (yellowness index); OSI (oxidative stability index, h).

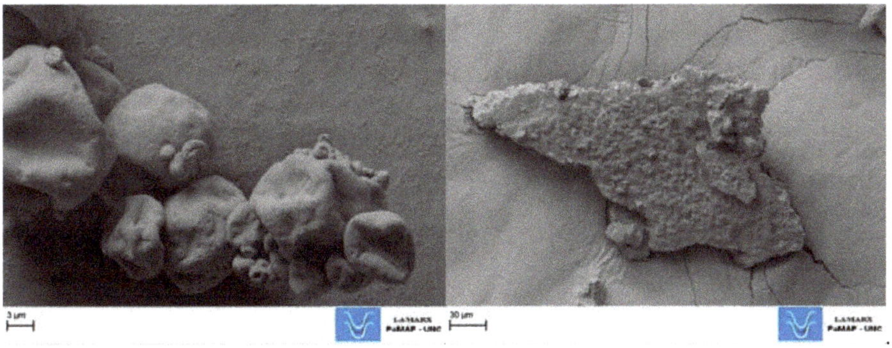

Figure 2. Scanning electron microscopy (SEM) images for microencapsulated CSO. (**Left**) spray drying; (**Right**) freeze drying.

As regards the flowing properties, CI > 25% and HR > 1.4 were observed, evidencing a poor flowability and strong inter-particle forces [11].

Significant changes in OSI values were observed ($p < 0.05$) according to the drying method and the interactions among biopolymers—from 6.45 to 12.04 h (freeze-drying) and from 12.05 to 15.31 h (spray drying) after inducing coacervation—which were possibly associated with the shifted denaturation temperatures (Table 1).

The porous, irregular and flake-like structure (Figure 2) of freeze-dried powders allows the diffusion of oxygen more rapidly than in spray-dried powders, which would explain the lower OSI values [1]. Nonetheless, higher OSI values than bulk CSO (3.00–3.33 h) were obtained with both drying methods.

4. Conclusions

The physico–chemical properties of microencapsulated CSO proved to be greatly affected by the drying method and the interactions among biopolymers in the ternary WPC/SPI/GA blend. Enhanced OSI values for microencapsulated CSO could be obtained by spray drying, through the formation of less porous solid microparticles compared with freeze-dried powders. In addition, even higher OSI values could be observed after coacervation in the ternary blend, possibly due to shifted WPC denaturation temperatures. It could be concluded that the ternary WPC/SPI/GA wall material blend constitutes an adequate matrix to encapsulate CSO.

Acknowledgments: This work was supported by grant IaValSe-Food-CYTED (Ref. 119RT0567), Fondo para la Investigación Científica y Tecnológica (FONCyT, BID PICT 2014-2283, and PICT 2016-1150) and SeCyT-UNC.

References

1. González, A.; Martínez, M.L.; Paredes, A.; Leon, A.; Ribotta, P.D. Study of the preparation process and variation of wall components in chia (*Salvia hispanica* L.) oil microencapsulation. *Powder Technol.* **2016**, *301*, 868–875, doi:10.1016/j.powtec.2016.07.026.
2. Timilsena, Y.; Adhikari, R.; Barrow, C.J.; Adhikari, B. Microencapsulation of chia seed oil using chia seed protein isolate □ chia seed gum complex coacervates. *Int. J. Boil. Macromol.* **2016**, *91*, 347–357, doi:10.1016/j.ijbiomac.2016.05.058.
3. Timilsena, Y.; Adhikari, R.; Barrow, C.J.; Adhikari, B. Digestion behaviour of chia seed oil encapsulated in chia seed protein-gum complex coacervates. *Food Hydrocoll.* **2017**, *66*, 71–81, doi:10.1016/j.foodhyd.2016.12.017.
4. Anandharanakrishnan, C.; Ishwarya, S. *Spray Drying Techniques for Food Ingredient Encapsulation*; John Wiley & Sons: Chichester, West Sussex, UK, 2015; ISBN 978-1-118-86419-7.
5. Burgos-Díaz, C.; Wandersleben, T.; Marqués, A.M.; Rubilar, M. Multilayer emulsions stabilized by vegetable proteins and polysaccharides. *Curr. Opin. Colloid Interface Sci.* **2016**, *25*, 51–57, doi:10.1016/j.cocis.2016.06.014.
6. Martínez, M.L.; Marín, M.A.; Faller, C.M.S.; Revol, J.; Penci, M.C.; Ribotta, P.D. Chia (*Salvia hispanica* L.) oil extraction: Study of processing parameters. *LWT Food Sci. Technol.* **2012**, *47*, 78–82, doi:10.1016/j.lwt.2011.12.032.
7. Canalis, M.S.B.; Valentinuzzi, M.C.; Acosta, R.H.; Leon, A.E.; Ribotta, P.D. Effects of Fat and Sugar on Dough and Biscuit Behaviours and their Relationship to Proton Mobility Characterized by TD-NMR. *Food Bioprocess Technol.* **2018**, *11*, 953–965, doi:10.1007/s11947-018-2063-z.
8. Us-Medina, U.; Julio, L.; Segura-Campos, M.R.; Ixtaina, V.Y.; Tomás, M.C. Development and characterization of spray-dried chia oil microcapsules using by-products from chia as wall material. *Powder Technol.* **2018**, *334*, 1–8, doi:10.1016/j.powtec.2018.04.060.
9. González, A.; Gastelú, G.; Barrera, G.N.; Ribotta, P.D.; Igarzabal, C.I.A. Preparation and characterization of soy protein films reinforced with cellulose nanofibers obtained from soybean by-products. *Food Hydrocoll.* **2019**, *89*, 758–764, doi:10.1016/j.foodhyd.2018.11.051.
10. Ma, T.; Zhao, H.; Wang, J.; Sun, B. Effect of processing conditions on the morphology and oxidative stability of lipid microcapsules during complex coacervation. *Food Hydrocoll.* **2019**, *87*, 637–643, doi:10.1016/j.foodhyd.2018.08.053.
11. Rodriguez, E.S.; Julio, L.; Henning, C.; Diehl, B.W.; Tomás, M.C.; Ixtaina, V.Y. Effect of natural antioxidants on the physicochemical properties and stability of freeze-dried microencapsulated chia seed oil. *J. Sci. Food Agric.* **2018**, *99*, 1682–1690, doi:10.1002/jsfa.9355.
12. Anandharamakrishnan, C.; Rielly, C.D.; Stapley, A.G. Effects of Process Variables on the Denaturation of Whey Proteins during Spray Drying. *Dry. Technol.* **2007**, *25*, 799–807, doi:10.1080/07373930701370175.

MDPI
St. Alban-Anlage 66
4052 Basel
Switzerland
Tel. +41 61 683 77 34
Fax +41 61 302 89 18
www.mdpi.com

Proceedings Editorial Office
E-mail: proceedings@mdpi.com
www.mdpi.com/journal/proceedings

www.ingramcontent.com/pod-product-compliance
Lightning Source LLC
LaVergne TN
LVHW070618100526
838202LV00012B/675